T0164129

MOGREB-EL-ACKSA

R. B. Cunninghame Graham

MOGREB-EL-ACKSA

A Journey in Morocco

Introduction by Edward Garnett

TMP

The Marlboro Press / Northwestern
Evanston, Illinois

The Marlboro Press/Northwestern
Northwestern University Press
Evanston, Illinois 60208-4210

Published by the Marlboro Press in 1985. The Marlboro Press/Northwestern
edition published 1997. All rights reserved.

Printed in the United States of America
10 9 8 7 6 5 4 3 2
ISBN 0-8101-6036-6

Library of Congress Cataloging-in-Publication Data

Cunninghame Graham, R. B. (Robert Bontine), 1852–1936.
   Mogreb-el-Acksa : a journey in Morocco / R. B. Cunninghame
Graham ; introduction by Edward Garnett.
      p.   cm. — (Marlboro travel)
   Originally published in 1898.
   ISBN 0-8101-6036-6 (alk. paper)
      1. Morocco—Description and travel. 2. Cunninghame Graham,
R. B. (Robert Bontine), 1852–1936—Journeys—Morocco. I. Title.
II. Series.
   DT309.G74 1997
   916.404'5—dc21   96-52696
                                                             CIP

To
HAJ MOHAMMED ES SWANI EL BAHRI
I Dedicate This Book

Not that he will ever read, or even, being informed of it, ever comprehend its nature, except in so far as to think it some "Shaitanieh" or another not to be understood.

But I do so because we have travelled much together, and far, and it must have been at times a sore temptation to him, in lonely places, to assure himself of Paradise by "nobbling" an unbeliever. Still, I would trust myself with him even to go the pilgrimage to Mecca; therefore, he must trust me when I swear not to have cast a spell on him (as Christians will upon occasion) by writing his name here for unbelieving men to wonder at.

"Show me Sohail and I will show you the moon."

Introduction

In writing on *Mogreb-el-Acksa*, December 9, 1898, Conrad speaks of the reader experiencing " a continuous feeling of delight: the persuasion that we have got hold of a good thing. This should work for material success." But he adds, " Yet who knows. No doubt it is too good."

It was. The book written as a record of a journey in Morocco and treating of Moroccan affairs *vis-à-vis* with Europe, incidentally challenged many social shibboleths of the day. The author had no particular thesis, but he generally derided the Victorian practice of putting the glass to the blind eye while declaring that it could see the blessings of western progress but not its blight. Now that the intelligent person is getting alarmed at the congestion of mechanized civilization and knows he is unable to escape from the tentacles of the octopus " progress," it is difficult to realize the irritation Cunninghame Graham's heterodoxy gave to the "leaders" of opinion thirty years back, to men " thinking imperially," to politicians, dons and business men and to the newspaper scribes. The book, hailed by the few in 1898 as an original masterpiece, was both too unorthodox and too witty for the serious British taste. Accordingly many years passed before it was reprinted, and I do not think that the second edition aroused any particular attention. In *Mogreb-el-Acksa* we have, however, Cunninghame Graham at his finest. The book is that rare thing, a spontaneous work of art. It is written with a verve and a brilliancy of tone that make it unique in English books of travel. Its subject, Moorish life and its semi-feudal habits and manners, is near to the author's heart; and his vision of this Oriental land is intimately woven with his many memories of the life and environment of similarly "backward" peoples, natives of other continents. The picturesque narrative, sharply detailed, marches gaily with a light foot though it carries a wealth of apt commentary. The incidents of travel and the many rencontres are lit up by the author's shrewd philosophy; and his sarcastic appreciation of the wise and the worldly, of knaves and fools and of all the pitfalls that await our poor humanity, combines with his gleaming irony to make a lively pattern on the page. It may be noted here that what often makes for a weakness in his Sketches, namely the breaking of the scene's atmospheric illusion by the irruption of his asides and comments, becomes in his travel book an artistic strength. For this Moroccan countryside he so brilliantly depicts is alive with a surprising variety of passing folk, also journeying to and fro amid the native tribesmen, and our traveller's searching eye is kept busy noting all the signs and tokens of their calling. So, whether it be the Kaid of Kintafi himself or his portly Chamberlain, or Sherifs, Jews, muleteers, negroes, herdsmen, wandering "talebs," Persian minstrels and acrobats whose chance intercourse cannot miss the houses, for the barking of the dogs will guide you, if it falls dark."

And then comes evening, and the travellers, still kicking at their horses' sides, straining their eyes, keep pushing forward, stumbling and objurgating on the trail.

R. B. CUNNINGHAME GRAHAM

GARTMORE, 1898.

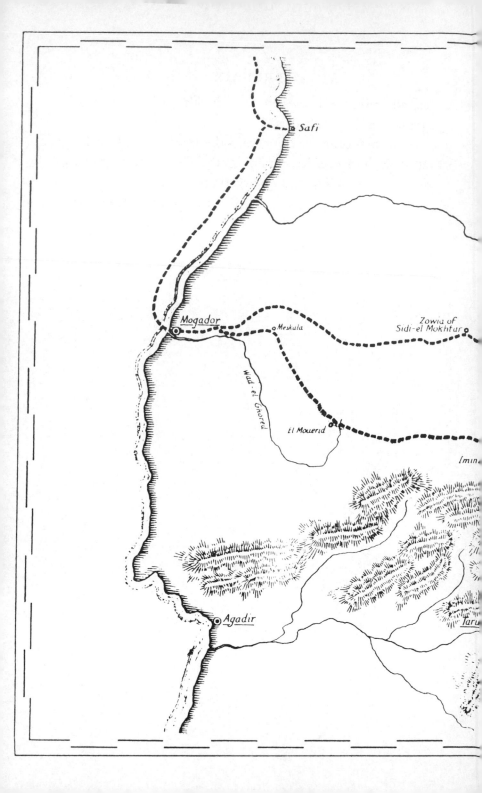

Chapter 1

SOUTH-SOUTH-WEST, a little westerly with a cloudy sky, and with long rollers setting into the harbour of Tangier, and the date October first in the last year of supposititious grace. The old white town, built round the bay with the Kasbah upon its hill, and the mosque tower just shining darkly against the electric light, for Tangier passed from darkness to electric light, no gas or oil lamps intervening, gave the effect as of a gleaming horse-shoe in the dark. Outside the harbour, moored twice as far away as the French ironclad, swung the *Rabat* at anchor, for the captain, a Catalan of the true Reus breed, was bound to supplement his lack of seamanship with extra caution. Towards her we made our way late in the evening in a boat manned by four strong-backed Jews, and seated silently, for it was the first stage of a journey concerning which each of our friends had added his wisest word of disencouragement. Nothing so spurs a man upon a journey as the cautions of his friends: "dangerous," "impossible," "when you get there nothing worth seeing," and the like, all show you plainly that the thing is worth the venture, for you know the world is ever proud to greet the conqueror with praise and flowers, but he has

3

to conquer first. And if he fails, the cautious kindly friend wags his wise head and shakes his moralizing finger, thinks you a fool, says so behind your back, but cannot moralize away experience, that chiefest recompense of every traveller. So in the boat besides myself sat Hassan Suleiman Lutaif, a Syrian gentleman, who acted as interpreter, and Haj Mohammed es Swani, a Moor of the Riff pirate breed, short, strong, black-bearded, with a turned-up moustache, and speaking Spanish after the Arab fashion, that is without the particles, and substitution of the gerund for every portion of the verb. El Haj and I were old acquaintances, having made several journeys in Morocco, and being well accustomed to each other's ways and idiosyncrasies upon the road.

Our bourne was Tarudant, a city in the province of the Sus, but rarely visited by Europeans, and of which no definite account exists by any traveller of repute. Only some hundred and fifty miles from Mogador, it yet continues almost untouched, the only Moorish city to which an air of mystery clings, and it remains the only place beyond the Atlas to the south in which the Sultan has a vestige of authority.*

Outside its walls the tribesmen live the old Arab life, all going armed, constantly fighting, each man's hand against his neighbour, and, therefore, in a measure, raised against

---

* Since writing this it has been shown that he has more authority in the Sus than was thought, as witness the landing of the Globe Venture Company's filibusters, which was promptly checked.

himself. And yet a land of vines, of orange gardens, olive yards, plantations of pomegranates, Roman remains, rich mines; but cursed " with too much powder," as the Arabs say, and therefore doomed, up to the present day, to languish without pauperism, prostitution, and the modern vices which, in more favoured lands, have long replaced the old-world vices which man brought with him into the old world. Even my friends were all agreed that to reach Tarudant in European clothes was quite impossible. Thus a disguise became imperative. After a long discussion I determined to impersonate a Turkish doctor travelling with his " taleb "*—that is, scribe—to see the world and write his travels in a book.† God, the great doctor, but under Him His earthly vicegerents, practicants, practitioners, with or without diploma, throughout the east, enjoy considerable respect; they kill, they cure, and still God has the praise. No one asks where they studied, and if faith in his powers helps a doctor in his trade, the east of all lands should be most congenial to all those who live by lancet, purge, and human faith. My stock of medicines was of the most homeric type, quinine and mercury, some Seidlitz powders, eye wash for ophthalmia, almost, in fact, as simple as that of the old Scotch doctor who doctored

---

* " Taleb " literally means a learned man, but in Morocco it is applied to everyone who can read and write, in fact, as the Spaniards say, " sabe de pluma " ; my " taleb " did not fall under the latter head, being a learned man, and a gentleman.

† I had to give up this as I spoke little Arabic and no Turkish, and as I looked rather like a Moor from Fez, finally called myself Sheikh Mohammed el Fasi; but I fear few were taken in by that name.

with what he styled the two simples, that is, laudanum and calomel.

The multitudinous dialects of Arabic, the constant travelling, either upon the pilgrimage or for the love of travel, which seems inherent in the Arab, render it possible for a European occasionally to pass unrecognized, even although his stock of Arabic is as exiguous as was my own. Behold us, then, approaching the *Rabat,* a steamer of the Linea Tras-Atlantica Española, bound for Mogador, some five days' journey below Tangier, the point at which I intended to put on Moorish clothes, buy mules and horses, engage a guide, and set out for Tarudant, the city where La Caba, Count Julian's daughter (she whose beauty tempted Don Rodrigo, and to avenge whose honour her father sold Spain to the Moors), lies buried, though how she came to die there is to historians unknown. On board the ship there was no light, no look-out, and the accommodation ladder was triced up for the night, stewards were asleep, and the long narrow vessel dipped her nose into the rollers, rolling and heaving in the dark, whilst we in Spanish, English, Arabic, and Portuguese rang all the changes upon " Ship ahoy." At last a sulky, sleepy quartermaster deigned to let down the ladder, and we scrambled upon deck.

Next morning found us off Arzila, the Julia Constantia Zilis of the Romans, a small, walled town lost among orange gardens, and girt by walls built by the Portuguese. Over the gateway the half-obliterated arms of Portugal can still be seen, but mouldering to decay, as is the country that they represent. Close to the walls rises the tomb of the Saint of Arzila; the white dome springs from a dense thicket of palmetto, dwarf rhododendron, berberis, and aromatic shrubs, a tiny stream bordered by oleanders (bitter as is an oleander is a proverb in the land) runs past it; above the dome and red-tiled roof two palms keep watch, and whisper in the wind the secrets palm trees know, and tell each other of the East, and of the various changes, Arab and Berber, Moor and Portuguese, they have seen during the course of their slow growth and long extended days. Columbus visited the place; the battle of Trafalgar was fought a few leagues off, and an old shepherd, still living where he was born, above the lighthouse on Cape Spartel, remembers to have seen it as a boy when he lay out upon the hills tending his goats.

Inside the town decay and ruin, heaps of garbage, a palace falling here, a hovel there. A town almost entirely given up to Jews, who carry on their business fleecing the Moors, and still complaining when at last the cent. per cented victim turns and rends. The best-known citizen is Mr. Ben Chiton, an Israelite of the best type, old, and white bearded, and with round, beady eyes, like an old rat, consul of seven nations, with all the seven flag-staffs on his hospitable roof. Dressed in the Jewish gown girt round the waist, wearing his slippers over white cotton stockings, and the black cup badge of the servitude of all his race throughout Morocco. His house is neat and clean, almost

beyond the power of endurance, Rahel and Mordejai labouring incessantly with broom and whitewash brush, and
he himself talking incessantly in the Toledan* dialect and
thick Jewish accent, of politics, of kings, ambassadors, and
as to whether England and France will go to war over
the question of the Niger and the like, whilst all the time
he presses brandy on you in a lordly glass holding about
a pint, which you must swallow no matter what the hour,
the while he calls upon the God of Isaac to bear him witness that the house is yours, and shows you, with just pride,
the chair on which he says the Prince of Wales once sat,
although historians do not seem to chronicle his passage
through the town. An Israelite of Israelites, a worthy Jew,
come of the tribe of Judah as he says, one to whose house
all travellers are welcome; loved by his family with ven-
eration, as the heads of Jewish houses are. Long may he
prosper, and may his roof grow broad as a phylactery, and
become strong enough to bear the flag-staffs of all the
nations upon earth, and his house long remain to show the
curious what most probably a Jewish house in Spain was
like before the Catholic Kings, in their consuming thirst
for unity, expelled the Jews, sweeping at one fell blow
commerce and usury out of their sacred land, and setting

---

* The Toledan dialect is the old Spanish of the Jews, who were expelled from
Spain. They speak the Spanish of the sixteenth century, using the X instead of J,
and with other peculiarities. They inhabit the towns of Tetuan, Tangier, Alcazar,
Arzila, Larache, and Rabat, and are the highest caste Jews of Morocco. The Jews
of the interior speak Arabic, come from the East, and are in general of a lower
type.

miserable Jew, fearfully sea-sick, balancing up in the gun-wale of a boat, in a mixed jargon of Arabic and Portuguese kept tale after a fashion, of the melons, and at last the vessel put to sea amid the curses of the passengers, and having earned the name amongst the Arabs of Abu Batigh, the Father of the Melons. Amongst the Arabs almost every man is Father of something or some quality, and lucky he who does not find himself styled, Father of the " ginger beard," or of " bad breath," or any other personal or moral failing, peculiarity, or notable defect.

Once more we urge our nine-knot course, and now find time to observe our fellow-passengers. In the cabin a Ger-man lady and her daughters are enduring agonies of sea-sickness, on the way to join her husband at Rabat.* The husband, known as the " mojandis," that is, engineer to the Sultan, proves, when we meet him, to be a cheery poly-glot blasphemer, charged with the erection of some forts at the entrance to the harbour of Rabat. For a wonder the bar of the river Bu-Regeh proves passable, and the German lady and her daughters can land, which is, it seems, a piece of luck, for the bar is known as the most dangerous of all the coast. Rabat, perhaps the most picturesquely situated of any city in Morocco, stands on a hill underneath which the river runs, and the spray from the bar is drifted occa-

* The Germans are slowly but surely driving us out of the coast trade in Morocco. It may well be said of them that they are conquering the world with the trombone and the ophicleide, as the Romans did with the pick and shovel. Where these " bummel bands " once get a footing, their empire and all that, is sure to follow.

sionally into the houses like a shower of snow. Here is the richest colony of the Spanish Jews, and here the best Morisco* families took refuge after the expulsion of the Moors from Spain. The town is estimated to contain some 20,000 inhabitants, and it is one of the four official capitals†; the Sultan has a palace with enormous gardens on the outskirts of the town. Just opposite is built Salee, called by the Moors " Salá," famous for having given its name to the most enterprising of the pirates of the coast in days gone by. Today it is a little white, mouldering place, baked in the fierce sun, or swept by south-west gales, according to the season and the time of year. The inhabitants are still renowned for their fanaticism, and the traveller who passes through the place is seldom able to dismount, but traverses the place trying to look as dignified as possible amid a shower of curses, a sporadic curse or two, and some saliva if he ventures within range of mouth. Many a poor Christian has worked his life out in the construction of the walls, and the superior whiteness of the inhabitants would seem to show that some of the Christian dogs have left their blood amongst the people of the place. Men still alive can tell of what a scourge the pirates were, and I myself once knew a venerable lady who in her youth had the distinction of having been taken by a Salee rover; an honour in

* The Moriscos went to Tetuan, Fez, and Rabat, and in consequence the upper classes of all these cities are fairer in colour and more enlightened in mind than those of any other city of Morocco.

† The others are Fez, Marakesh, and Mequinez. The system of an ambulatory capital also obtained formerly in Bolivia.

itself to be compared alone to that enjoyed by ladies who are styled peeresses in their own right. Robinson Crusoe, I think, once landed at the place, and the voyages of imaginary travellers make a place more authentic than the visit of the unsubstantial personages of real life. A little up the river is the deserted city of Schellah, and on the way to it, upon a promontory jutting into the stream, is built the half-finished tower of Hassan, an enormous structure much like the Giralda at Seville, and by tradition said to be built by the same architect who built that tower, and also built the tower of the Kutubieh, which, at Morocco city, serves as a landmark in the great plain around. The Giralda and the Kutubieh spring from the level of the street, but the Hassan tower excels them both in site, standing as it does upon a cliff, and " looming lofty "* as one passes in a boat beneath. Schellah contains the tombs of the Sultans of the Beni Merini dynasty. El Mansur (the victorious) sleeps underneath a carved stone tomb, over which date palms rustle, and by which a little stream sings a perpetual dirge. The tomb is held so sacred that till but lately neither Christians nor Jews could visit it. Even today the incursions of the fierce Zimouri, a Berber tribe, render a visit at times precarious. The walls of the town enclose a space of about a mile circumference; sheep, goats, and camels feed inside them, and a footpath leads from one deserted gate-house

---

* Once looking at a steamer coming up London docks in a thick fog, I said to an aged shell-back standing near, " She looks very large somehow," and received the answer, " Well, she do loom lofty in the mist."

to the other, a shepherd boy or two play on their reeds, and though the sun beats fiercely on the open space, it looks forlorn and melancholy, and even the green lizards peering from the walls look about timidly, as if they feared to see a ghost. On gates and walls, on ruined tombs and palaces, the lichens grow red after the fashion of hot countries, and the fine stonework, resembling the stucco work of the Alhambra, remains as keen in edge and execution as when the last stroke of the chisel turned it out. Outside the town are olive yards and orange gardens, and one comes upon the long-deserted place with the same feelings as a traveller sees Palenque burst on him out of the forests of Yucatan, or as in Paraguay after a weary following of dark forest trails, the spires of some old Jesuit Mission suddenly appear in a green clearing, as at Jesus or Trinidad, San Cosme Los Apostoles, or any other of the ruined " capillas," where the bellbird calls amongst the trees, and the inhabitants take off their hats at sunset and sink upon their knees, bearing in their minds the teaching of the Jesuits, whom Charles III expelled from Spain and from the Indies, to show his liberalism.

Our most important passenger was Don Jose Miravent, the Spanish Consul at Mogador, returning to his post after a holiday; a formal Spaniard of the old school, pompous and kind, able to bombast out a platitude with the air of a philosopher communicating what he supposes truth. At dinner he would square his elbows, and, throwing back his head, inform the world " the Kings of Portugal are now at Caldas"; or if asked about the war in Cuba, say: "War, sir, there is none; true some negroes in the Manigua* are giving trouble to our troops, but General Blanco is about to go, and all will soon be over; it is really nothing (no es na), peace will, please God, soon reign upon that lovely land." At night he would sit talking to his cook, a Spanish woman (the widow of an English soldier), whom he was taking back to Mogador; but though he talked, and she replied for hours most volubly, not for a moment were good manners set aside, nor did the cook presume in any way, and throughout the conversation talked better than most ladies; but the "tertulia" over, straight became a cook again, brought him his tea, calling him "Amo" (master), "Don Jose," and he quite affably listened to her opinions and ideas of things, which seemed at least as valuable as were his own. A Jewish merchant dressed in the height of Cadiz fashion, and known as Tagir† Isaac, occupied much the same sort of social state on board as does a Eurasian in Calcutta or Bombay. Tall, thin, and up-to-date, he had divested himself almost entirely of the guttural Toledan accent, but the sign of his "election" still remained about his hair, which tended to come off in patches like an old hair trunk, and at the ends showed knotty, as if he suffered from the disease to which so many of his

---

* Manigua is the Cuban name for a tropical forest.

† Tagir means a merchant and is a highly honourable title in Mohammedan countries where the feudal system never made trade be looked down upon. "Caballer," i.e. Caballero, is a more modern and, so to speak, gentleman-like appellation.

compatriots in Morocco are subject, and which makes each separate hair stand out as if it were alive.*

But, as is often seen even in more ambitious vessels than the *Rabat*, the passengers of greatest interest were in the steerage. Not that there was a regular steerage as in ocean-going ships, but still, some thirty people went as deck passengers, and amongst them was to be seen a perfect microcosm of the Eastern world. Firstly a miserable, pale-faced Frenchman, dressed in a dirty " duster " coat, bed-ticking "pants," black velvet waistcoat, and blue velvet slippers, with foxes' heads embroidered on them in yellow beads, his beard trimmed to a point, in what is termed (I think) Elizabethan fashion, and thin white hands, more disagreeable in appearance than if they had been soiled by all the meanest work on earth. His "taifa"† (that is band) consisted of one Spanish girl and two half-French half-Spanish women, whom he referred to as his "company," and whom he said were to enact "cuadros vivos," that is " living pictures," in the various ports. They, less polite than he, called him " el Alcahuete," which word I leave in Spanish, merely premising (as North Britons say) that it is taken from the Arabic " el Cahueit," and that the celebrated

* Perhaps a kind of Plica Polonica, or perhaps the " scald " of the Middle Ages, as when Chaucer says to his scribe who did not correct his proofs, "Under they long lockes mayst thou have the scald."

† Taifa is the Arabic word for a company or following. It is generally applied contemptuously, as in the phrase "Reyezuelos de Taifas," used by the Spaniards during the time the Moors were in Spain to designate the Kings of Almeria, Ocsonoba, Huesca, and other small Principalities.

"Celestina"* was perhaps in modern times the finest specimen of the profession in any literature. So whilst our captain read "Jack el Ripero" (it cost him two pesetas when
new in Cadiz), I take a glance at the inferior races, who
were well represented in the steerage of the ship.

Firstly, I came across a confrère in the healing art who
hailed from Tunis, a fattish Arab dressed in Algerian
trousers, short zouave jacket, red fez, pink and white sash,
and watch chain of two carat silver with an imitation seal
of glass, brown thread stockings and cheap sand shoes after
the pattern to be seen at Margate. "Tabib numero Wahed,"
that is an A-1 doctor, so he says, has studied in Stamboul,
where he remained nine years studying and diligently
noting down all that he learnt. But, curiously enough, all
that he now remembers of the Turkish language is the
word "Mashallah," which he displayed to my bewilder-
ment, until Lutaif, who spoke good Turkish, turned on
a flood of pertinent remarks, to which the doctor answered
in vile Tunisian Arabic, not having understood a word.
Though not a linguist, still a competent practitioner, tre-
panning people's heads with ease, and putting in pieces of
gourd instead of silver as being lighter, and if the patient
died, or by the force of constitution lived through his
treatment, the praise was God's, Allah the great Tabib
(doctor), although he sends his delegates to practise on
mankind, just in the same way that in the East each man

* Celestina, Tragi-Comedia de Calisto y Melibea por Rodrigo de Cota, Juan
de Mera or some one else unknown.

commits his work to some one else to do. At least he said so, and I agree, but fail to see the use of substituting gourd for silver, seeing the vast majority of heads are gourds from birth. Quinine he had, and blistering fluid, with calomel and other simples, and when the Christian quinine ran out, he made more for himself out of the ashes of an oleander stick mixed with burnt scales of fish and dead men's bones, and found his preparation, which he styled " El quina beladia " (native-made quinine), even more efficacious than the drug from over sea. Also he used the seven herbs, known as the " confirmed herbs," for tertian fevers, with notable success. A cheerful, not too superstitious son of Æsculapius, taking himself as seriously as if he had had a large brass plate upon his door in Harley Street, and welcomed by his surviving patients in Rabat on our arrival with great enthusiasm.* His wife, a shrouded figure lost in white veils and fleecy shawls, had the dejected air that doctors' wives often appear to labour under, even in countries where they can rule their husbands openly.

A group of high-class Arabs sat by themselves upon the decks, waited upon by a tall tribesman faced like a camel, and with the handle of his pistol and curved dagger outlined beneath his clothes. This group was composed of sherifs† from Algeria, all high-caste men, dignified, slow,

---

* This may serve to illustrate the retrograde condition of medicine in Mohammedan countries. Who would welcome the return of a doctor in a European country?

† A sherif is a more or less authentic descendant of the Prophet. They occupy a semi-religious semi-political position, and are as numerous as the " ancestors came over with the Conqueror " people in England.

and soft of speech, deliberate in movement, their clothes as white as snow, nails dyed with henna, each with a heavy rosary in his hand, their business somewhat mysterious, but bound to see the Sultan in his camp. Throughout Algeria the sherifs are not allowed to levy contributions from the people openly, but it is said in private they receive them all the same. Of all the population they are the least contented with French rule, as since the conquest naturally they have fallen somewhat from the position they once occupied, and cannot go about receiving presents for the pains they have taken to preserve their lineage, as they do in Morocco, making themselves a travelling offertory.

All of them wished to know of the late war between the Sultan of Turkey and the " Emperor of the Greeks." They seemed to think the latter was a descendant of the *Paleologi*, and asked if it were true the Sultan had killed all the Greeks except fifteen, and if these latter had not fled to " Windres " (London) to seek protection from the Great Queen and to advise her to make preparations against the " Jehad "* to come. With them they carried sundry hide bags of gold with which they said they wished to purchase permission from the Sultan of Morocco to export grain, as the harvest in Algeria had not been good owing to locusts, and the lack of rain. Of course this may have been the object of their visit, but since the Greco-Turkish war all the Mohammedan world is on the stir, and men

---

* Jehad—religious war—generally applied to a war entered into from self-interest, as that of the United States against Spain.

are travelling about from place to place disseminating news, and all the talk is on the victories of the Turks, and on the rising of the tribes in India; in fact, a feeling seems to be abroad that the Christian power is on the wane, and that their own religion once again may triumph and prevail.

At Casablanca—called by the Arabs Dar el Baida, that is, the White House—the sherifs go ashore, and I last saw them seated on their bags, outside the waterport, their backs against the ramparts of the town, their eyes apparently fixed upon nothing though seeing everything, telling their beads and waiting patiently, enduring sun and flies until their servant should return and tell them of a lodging fit for persons of their quality. Of all the towns on the Morocco coast Dar el Baida has the most business, the country at the back of it is fertile and grows much wheat, the tribes are fairly prosperous, and the best horses of the country come from the districts known as Abda and Dukala, a few leagues from the place.

Consuls abound, of course, so do hyenas—that is, outside the town—but both are harmless and furnish little sport, except the Consul of America, my good friend Captain Cobb. He, if my memory fails me not, piled up his brig some thirty years ago upon the beach in the vicinity, liked the climate, became a Consul, naturally, and to this day has never returned to his sorrowing family in Portland, Maine.

In thirty years tradition says he has not learned a word

of Arabic with the exception of the word "Balak" (look out), which he pronounces "Balaaker," and yet holds conversations by the hour in Arabic, and both the patients seem contented with their lot. All the attractions to be met with in the town do not detain me; what takes my fancy most is to see tribesmen from the country, armed to the teeth, and balancing a gun full six feet long upon their saddles, sit on their horses bargaining at shops in the same fashion I have seen the Gauchos at a camp-town in South America, their horses nodding their heads and looking half asleep, their owners seated with one leg passed round the pommel of the saddle, and passing hours seated as comfortably as in a chair.

Back to the steamer in a boat, and at the waterport we pass a group of Jews washing themselves, in preparation for a feast. Lutaif ranks as a wit for saying that the Jews will defile the sea, for any wit is small enough to bait a Jew with; and the Arabs, though they will say the same things of a Christian behind his back, all laugh consumedly when a Christian takes their side against a Jew.

On board the uneasy ship, tossing like a buck-jumper in the Atlantic swell, we find more Oriental items ready to hand. The first a tall, thin, cuckoldy-looking Arab knave, dressed in a suit of slop-made European clothes, his trousers half-a-foot too short, his boots unblacked, and himself closely watched by two Franciscan Friars.* It appeared he

* For the last two hundred years the Franciscans have had missions in the coast towns of Morocco. They, at present, confine their labours to the Spanish popula-

was a convert. Now, in Morocco a convert is a most rare
and curious animal, and he is usually not a great credit to
his capturers. On this occasion, it appears, the convert had
been dallying with the Protestants, had given them hopes,
had led them on, and at the last, perhaps because he found
the North-British water* of their baptism too cold for him,
or perchance because the Friars gave a dollar more, had
fallen away to Rome. However, there he was, a veritable
" brand," a sheep who had come into one of the folds,
leaving the other seven million nine hundred and ninety-
nine thousand and ninety-nine still straying about Morocco,
steeped in the errors of Mohammedanism. His captors were
a gentlemanlike, extremely handsome, quiet Castilian, who
to speak silver (hablar en pata) seemed a little diffident
about his prize, and went about after the fashion of a boy
in Texas who has caught a skunk. The other guardian
had no doubts. He was a sturdy Catalonian lay-brother,
who pointed to the " brand " with pride, and told us, with
a phrase verging upon an oath, that he was glad the Protestants had had their noses well put out of joint. The

tion. The vineyard is stony. Formerly they worked amongst the captives. Your captive generally is inclined to listen to a friar or to any one else who will talk to him.
The late Prefect of the Franciscans in Tangier, Father Lerchundi, was a most
erudite man, having composed many treatises on the Arabic language, which he
knew perfectly. At his funeral, Jews, Christians, Moors, and other mutually
described infidels, turned out in great numbers, and for a brief space the
" Odium Theologicum " was laid aside, and the cross, the crescent, and the other
symbols of the three jarring faiths went up the main street of Tangier in seeming
amity.

  * The Protestant missionaries in Morocco are almost all Scotchmen. I have
received unfailing kindness from all of them, whether as a Protestant or a Scotchman, I do not know.

victim was a merry sort of knave, who chewed tobacco, spoke almost every language in the world, had travelled, and informed me, when I asked him where he was going, that the "Frayliehs" (Friars) were taking him to Cadiz "to have the water put upon his head." He seemed an old hand at the business, and recognized my follower, Swani, as a friend, and they retired to talk things over, with the result that, ere night fell, the "convert" was in a most unseemly state and singing Spanish songs in which Dolores, Mercedillas, and other "Chicas" figured largely, and were addressed in terms sufficient to upset a convent of Franciscan Friars. Peace to his baptism, and may the Protestants, when their turn comes to mark a sheep, secure as fine a specimen as the one I saw going to Cadiz to have "the water put upon his head." This missionary question and the decoying of God-fearing men out of the ranks of the religion they were born in, is most thorny in every country like Morocco, where the religion of the land is one to which the people are attached. An earnest missionary, a pious publican, a minister of the crown who never told a lie, are men to praise God for, continually. Honour to all of them, labouring in their vocations and striving after truth as it appears to them. I, for my part, have found honest and earnest men both in the Scottish missions in Morocco and in the ranks of the Franciscan Friars from Spain. Amongst both classes, and in the missions sent by other churches, good men abound; and in so far as these good men confine themselves to giving medicines, healing the sick, and showing by the example of their lives that even Christians (whom Arabs all believe are influenced in all they do by money) can live pure, self-denying lives for an idea, the good they do is great; but that by living, as they do, amongst the Moors, they do more good than they could do at home by living the same lives, that I deny. Amongst Mohammedans plenty of people lead good lives, as good appears to them—that is, they follow out the precepts of their faith, give to the poor, do not lend money upon usury; and, to be brief, practise morality,* and believe by doing so that they are sacrificing to some fetish, invented by mankind to make men miserable. When, though, it comes to marking sheep—the object, after all, for which a missionary is sent—I never saw a statement of accounts which brought the balance out upon the credit side. The excuse is, generally, " Oh, give us time; these things work slowly "—as indeed they do; and if the missions think it worth their while to send men out to doctor syphilis, cure gonorrhœa, and to attend to every form of the venereal disease, their field is wide; but if they wish to convert Mohammedans let them produce a balance-sheet, and show how many of the infidel they have converted in the last twenty years. Not that I blame them for endeavouring to perform what seems impossible, or try to detract an atom from the praise due to them for their efforts, but when so many savages still exist in Central

* Morality from *mores*, customs; therefore, as the customs of all nations are different, so is their theory of morality.

Africa and in East London—not to speak of Glasgow and
the like—it seems a pity to expend upon a people, civilized
according to their needs, so much good faith, which might
be used with good effect upon less stony ground.

Prophets, reformers, missionaries, "illuminated" folk,
and those who leave their homes to preach a faith, no
matter what it is, are people set apart from the flat-footed
ordinary race of human kind; of such are missionaries and
the dream world they live in. How many an honest, hard-
working young man, who gets his education how he can,
by pinching, screwing, reading at bookstalls, paying for
instruction to those who live perhaps almost as poorly as
he himself, thinks when he reads of Livingstone, Francisco
Xavier, the Jesuits of Paraguay, and Father Damien, that
he, too, would like to take his cross upon his back and
follow them. It seems so fine (and is so splendid in reality)
a self-denying life, lived far away from comforts, without
books; Bible and gun in hand, to show the heathen all the
glories of our faith.

Then comes reality, mocker of all best impulses, and the
enthusiastic spirit finds itself bound in a surgery (say in
Morocco city), finds work increasing, and his dreams of
preaching to a crowd of dusky catechumens dressed in
white, with flowers in their hair, and innocence in every
heart, turned to Dead Sea Fruit, as with motives misunderstood, with caustic, mercury, sulphate of copper and
the rest, he burns the fetid ulcer, or washes sores, and for
reward can say after a term of years that he has made

the people of the place look upon European clothes with less aversion than when first he came.

If the mere fact of getting accustomed to the sight of Europeans were a thing worthy of self-congratulation, then indeed missionaries in Morocco have achieved much. But as I see the matter Europeans are a curse throughout the East. What do they bring worth bringing, as a general rule?

Guns, gin, powder, and shoddy cloths, dishonest dealing only too frequently, and flimsy manufactures which displace the fabrics woven by the women; new wants, new ways, and discontent with what they know, and no attempt to teach a proper comprehension of what they introduce; these are the blessings Europeans take to Eastern lands. Example certainly they do set, for ask a native what he thinks of us, and if he has the chance to answer without fear, 'tis ten to one he says, Christian and cheat are terms synonymous. Who that has lived in Arab countries, and does not know that fear, and fear alone, makes the position of the Christian tolerable. Christ and Mohammed never will be friends; their teaching, lives, and the conditions of the different peoples amongst whom they preached make it impossible; even the truce they keep is from the teeth outwards, and their respective followers misunderstand each other quite as thoroughly as when a thousand years ago they came across each other's path for the first time. But if the Arabs constitute a stony vineyard, the Jews are worse, and years ago when first the missionaries appeared in the coast towns of southern Barbary, they fleshed their maiden weapons on the Jews. It struck the chosen people that the best weapon to employ against their new tormentors was that of irony, and so they cast about to find a nickname calculated at the same time to ridicule and wound, and found it, made it stick and rankle, so that today every new missionary on landing has to accommodate his shoulders to the burden of a peculiarly comic cross.

Almost all Europeans in Morocco must of necessity be merchants, if not they must be consuls, for there is hardly any other industry open to them to choose. The missionaries bought and sold nothing, they were not consuls; still they ate and drank, lived in good houses, and though not rich, yet passed their lives in what the Jews called luxury. So they agreed to call them followers of Epicurus, for, as they said, " this Epicurus was a devil who did naught but eat and drink." The nickname stuck, and changed into " Bikouros " by the Moors, who thought it was a title of respect, became the name throughout Morocco for a missionary. One asks as naturally for the house of Epicurus on coming to a town as one asks for the " Checquers " or the " Bells " in rural England. Are you " Bikouros "? says a Moor, and thinks he does you honour by the inquiry; but the recipients of the name are fit to burst when they reflect on their laborious days spent in the surgery, their sowing seed upon the marble quarries of the people's hearts, and that the Jews in their malignity should charge upon them by this cursed name, that they live in Morocco to escape hard

work, and pass their time in eating and in quaffing healths
a thousand fathoms deep.

Often at night, awake and looking at the stars and trying
to remember which was which, I have broken into laughter
when I thought upon the name, and laughed until the
Arabs all sat up alarmed, for he who laughs at night laughs
at his hidden wickedness, or else because a devil has possessed his soul.

Cargo, of course, was not expected in the *Rabat,* and as
by four o'clock in the afternoon the last few passengers had
come aboard, it might have been expected that the ship
would put to sea. Not so, the captain gave it as his opinion
that it was best to wait till midnight so as to arrive in
Mazagan by breakfast time; so for eight hours we waited
in sight of Casablanca, swinging up and down, giddy and
miserable, and in the intervals of misery tried to fish.

Next morning we were off Djedida (New Town), as
the Arabs call Mazagan. A yellowish town, peopled by
Jews, Moors, and innumerable yellow dogs. Camels blocked
all the streets, and above the consul's flagstaffs a single
palm-tree reared its head, fluttering its feathers over the
sandstone walls built by the Portuguese in the days when
they were navigators, adventurers, and over-ran the world,
after the fashion of the modern Englishman. The batteries
built by the Moors or by the Portuguese are most ingen-
iously constructed to expose the gunners, and to batter
down the town they are supposed to guard. Outside a street
of beehive-looking huts of reeds, each with its little garden,

its ten or twelve dogs, thirteen · fourteen children, three
or four donkeys, and a score of mangy fowls, and with a
bush or two of castor oil plants sticking up in the sand
at every corner of the street, gives quite an air of equatorial
Africa or Paraguay.

Right in the middle of the Plaza a squadron of some
fifty Arab irregular soldiers guard about two hundred
prisoners. These latter, heavily ironed and half-starved, were
of the tribe of the Rahamna, who for the past twelve
months had been in open rebellion against their liege lord
the Sultan, Abdul Aziz, whom may God defend. The system in Morocco when a Sultan dies is for the tribes who
think they may be strong enough, to refuse to pay all taxes,
and then the Sultan is obliged to come in person with an
army to discuss the matter, to fight, and if he wins send
baskets full of heads to be stuck up on the chief cities' gates,
and squads of prisoners like the band I saw in Mazagan
to rot in prison till death relieves them of their miserable
lives. Every one of the prisoners in the Plaza must have
known that any chance of his release was small; indeed,
they knew they were going to be half-starved, either in a
dark cellar-like prison, or in an open courtyard exposed to
sun and cold, for the Morocco Government allows no ra-
tions to its prisoners, and those who have no friends soon
die of hunger.* Still they were not apparently much put

---

* Shortly before we left Tangier, my interpreter, Mr. Lutaif, took a man out
of the prison who had been five days without food. His offence was the possession
of a good djellaba (hooded cloak) and fifteen dollars.

about. People stood talking to them, flies settled on their eyelids, dogs licked their sores, horses and camels jostled against them, the sun poured on their heads, and still they did not suffer in the way our prisoners suffer, for they were not cut off in sentiment from those, their brothers in Mohammed, who stood round and talked to them to pass the time. Pious old ladies, young philanthropists and novelists in search of " copy " who visit Tangier go into " visibilio," as the Italians say, over the prisoners in the prison, with its flat horseshoe arches at the door, so similar in style to those at Toledo in the old synagogue, now turned into a church.*

All those good people and each journalist, all have their word about the darkness, chains, and the want of air; the misery, the crowding at the door to take the chocolate (which should have been tobacco) which the pious lady brings. Each tourist sends home his badly written paragraph to his favourite paper, and goes away lamenting over the barbarism of Mohammedans; and, if he is a " glorious empire " man, thinks of the time when, under the Union Jack, the Moorish prisoner shall have a number and a cell, tin pan to wash in, Bible to read, and all shall be apparelled in Queen Victoria's livery, until they purge themselves of their contempt and promise to amend their naughty lives.

One thing they all forget when writing of a Moorish prison, that, in spite of dirt, of chains, of want of air, of herding all together in a den, they are happier than pris-

* Santa Maria la Blanca.

oners with us, for they can speak, exhale their misery in
conversation; they still are men, and leave the prison men,
instead of devils, hating all mankind like those who, under
our inhuman silent plan, eat out their miserable " terms "
cursing the fools who in their foolish kindness hit on a
device to turn men into stone.* And so I leave the prisoners
in the square, thinking as an old Arab says: " It looks as
if the Sultan wished to finish with all the Mussulmen." At
any rate, since the last Sultan's death, three years ago,
thousands must have been killed in war, died by starvation,
or rotted miserably to death in prisons underneath the
ground.

It being a feast day of the Jews all stores were closed
in Mazagan, for in Morocco all the retail business is in
Jews' hands, and no religion that I know of seems to have
so many days on which men may not work; and therefore
it is strange that men have not yet flocked to join their
faith.

At last the steamer shakes itself free from the allure-
ments of Mazagan and steers almost due west to clear the
reef, which, jutting out about six miles, makes Mazagan
the least exposed of all the open harbours on the coast.
As Mazagan is distant only three days from the city of
Morocco it may be destined some day to a glorious commercial future, with railways, docks, smoke, pauperism,

---

* Elizabeth Fry never ceased remonstrances against the silent and solitary plan.
In a communication sent to M. de Beranger, of Paris, she urges, under the 7th
head, " the impossibility of fitting the prisoners for returning to society under the
system."

prostitution in the streets, twenty-five faiths instead of one, drunkards, cabs, bicycles and all our vices, so different in their nature from those the Arabs brought from Arabia and have clung to in the same way they cling to their religion, dress, speech, their alphabet, and their peculiar mode of life.

As we steamed out, the town of Azimur was seen under our lee, about twelve miles from Mazagan, situated on a hill close to the river Um er R'bieh, once an important harbour but long silted up by sand. The harbour, once the resort of pirates, is now silted up, and the remembrance of the pirates' deeds is still kept fresh in people's memories by a great store of old Delft plates, either taken by pirates in the merry days gone by, or sent from Holland* in the times when vessels from the Dutch ports traded along the coast. Today the foreign merchants buy them or exchange them for modern china, and their inglorious end is to hang on a wall beside the so-called artistic plenishings of ladies' over-gimcracked drawing-rooms.

As we turn southwards, after clearing the long reef, it seems as if we had already entered another zone. The hills along the beach become more arid, the plains all stony or covered with low, thorny scrub. Saffi, the hottest place on all the coast, melts past, merely a film of white against the reddish background, and in the distance the foot-hills ris-

* It was from Azimur that one of the companions of Alvar Nuñez came, in his captivity in Florida, for he says, "el quarto era negro alarabe de Azimur, se llamaba Estebanico." The Moors call the place Mulai Bushaib.

ing to the Atlas now appear. The largest of them, Jebel Hadid, the Mount of Iron, standing out like an enormous box* above the coast.

A Yemeni Jew has come on board, bound for Mogador, the city of the Jews. Short, black-eyed, greasy-locked, and with a red fez bound round with a woman's shawl, he, like St. Paul, is of a mean appearance. Still he has much to tell about Arabia, and the province from which he comes. It appears that near Sannaar, in South Arabia, there is a land called Beni-Mousa. In this favoured spot the inhabitants arc all Jews, and none of them are known to speak untruths, at least so says their representative on board the ship. This tribe, he says, is that of Shebatat. A river, which bounds their country, has so fierce a current that rocks and stones are moved along by it, and no mere Gentile can cross the stream. Upon its banks grow two tall trees (Sandaracs, he thinks), which bend across to one another and salute by saying, " Shabat Sholom." On Saturday they do not bend over to one another, and keep no watch; therefore, on Saturday alone can Gentiles cross the wondrous river, as on that day only does the stream abate its force. All this is true, and I myself am much confirmed in my opinion of its truth, because at night this same veracious Jew produced out of a bag a bottle of spirit (majia) made from dates, and, drinking it, got most uproarious, shouting and singing, falling repeatedly upon the winch

---

* The Cofre de Perote, in Mexico, is also box-like in appearance, as the name indicates.

to the great delight of all the Moors, and towards midnight avowing his intention of swimming back to the Yemen so that my henchman Haj Mohammed es Swani, who had been a sailor, had to make him fast to a ringbolt in the deck. And as in Arabic without a particle, after the fashion of a child or negro, I tried to express my astonishment at such a line of conduct in so grave a man, to a young Arab who stood near me, without a smile he answered, in most perfect English, " I am not certain if I understand all that you say, sir, but I speak English pretty well." In outward visible appearance he was an Arab of the usual kind, bare legs and yellow slippers, shaved head and fez with rather grimy turban, dirty white drawers, and brown djellaba* with the hood of it over his head after the fashion of a friar. It seemed he was an acrobat from the province of the Sus, from the celebrated Zowia† of Si Hamed O'Musa, the patron saint of all the acrobats throughout Morocco. Not far from the Wad‡ Nun is situate the district called Taseruelt, and in that district the famous patron saint of acrobats is buried at the above named Zowia. From thence a large proportion of the Arab troops of acrobats, who perform at our music-halls, set forth to tumble round the world.

My English-speaking acrobat had terminated an engage-

---

* Djellaba is the hooded garment shaped like a sack, much worn in Morocco, and apparently of Berber origin, as it is unknown in other Arab countries.

† A Zowia is the house and district of some Sheikh or Sherif.

‡ Wad, in Morocco, means river. It is the same word as the Egyptian Wadi, a dry valley. It appears in many Spanish names of places, as Guadalquivir, Guadalimar, Guadarrama, etc.

ment at the South London Music Hall, and was returning to
re-steep himself in the true faith at home, and also, it is
possible, to rest. He told me that his ambition was to marry
a European girl, and that his choice would be a German, for
he said, " German girls mind the house and sew; English
are prettier, but will not sew, and, besides that, they are
always drinking plenty." Out of the mouths of babes and
heathens I think a reflection on our national femininity
comes with some force. As we stand talking of the
" Empire," " Pavilion," " Oxford " and other " halls," both
in the provinces and the metropolis, the island which half
shelters the roadstead of Mogador came into view.

Kissed by the north-east trade which just envelopes Mogador and about half a league of country outside the town, the
city, dazzlingly white, lies in the sun, well meriting the
name of Sueira, that is, The Picture, given by the Moors.
Founded in 1760, by order of the Sultan Sidi Mohammed,
the plan made by an engineer from France, whose name,
according to the Arabs, was Cornut, the city (supposed by
some to be the ancient Erythræa) is the most regularly built
and most commercial of all the empire. A little desert,
varying in breadth from three to thirteen miles, cuts off the
city from the fertile lands. Sand, and more sand, fine, white,
and almost always altering in position, gives an idea of the
Sahara made in miniature. The little river Wad el Ghoreb
runs near the town, and in the middle of the water a former
Sultan has built a palace founded on the sand; but though
the north-east trade blows almost all the year, and when it

rains the rain comes down in torrents, the palace has not fallen, and, as it never was inhabited, it remains a monument of human folly, surmounting all the powers of providence. Jews, Jews, and still more Jews possess the place; they make their Kidush, keep their Purim, Cabañas, and New Year; eat adafina, broaden their business and phylacteries, are hospitable, domestic in their habits, each man revered in his own house as if he were a prophet, and all the business of the place is in their hands.

Grave, reserved Jews in gaberdines, smart up-to-date young Jews in white straw hats and European clothes, daughters of Israel with handkerchiefs bound round their heads and hanging down their backs, others in Paris fashions, but all with hair like horses' tails, are everywhere. Donah and Zorah, Renia, Estrella, Rahel, and Zulica, with Azar, Slimo, Baruch, and Mordejai are seen in every street; they sit in shops, lean out of windows, lounge upon the beach, walk about slowly as if they stepped on eggs, are kind in private life, cruel in business; they keep up much communication with Houndsditch, Hamburg, Amsterdam, Jerusalem, and other centres of the " community," speaking an Arabic garnished with English, seasoned with Spanish, peppered with Shillah* words, and rendered as intelligible as Chinook by the thick utterance with which they speak. An amiable race, business once over, and charitable amongst

---

* Shluoch is the Arabic name for the Southern Berbers, *i.e.* of the Atlas and the Sahara; Shluoch, in Arabic, means " cast out," and the language is called Shillah, in Arabic. The Shluoch call themselves Amazirght, *i.e.* the noble people. This difference of opinion as to nomenclature has been observed in other nations.

themselves; what is called moral, that is, their customs do not tolerate prostitution, and husbands love their wives, children their parents, and their home life resembles that which writers say is to be seen in England, but which experience generally shows is oftener found in France, where families go out together and men are not ashamed to play with children, and to sit drinking coffee out of doors beside their grandmothers.

So to this New Jerusalem, after a five days' voyage, the *Rabat* arrived, and anchored inside the island where the Sultan has a great open prison; there his rebels are confined, and daily die, both those who live and those whom death releases. An iron steamboat, much like a tramp in shape, but armed with four small guns, and commanded by a German officer, displays the red flag of Morocco, identical in colour with the well-known flag which in Hyde Park has braved a thousand meetings, and around which the " comrades " flock to listen when Quelch holds forth on social wrongs, or Mr. Hyndman speaks on India, and outside the crowd the bourgeois feels a shrinking in the stomach as he smooths his hat to give him countenance. We bid the microcosmic craft good-bye, and go ashore stuffed in a boat with Moors and Jews, some Spaniards, two Franciscan friars, eight or ten bird-cages, and land to find the Jewish feast proceeding, the hotels all full, the shops all shut, and the whole town delivered over to the mercies of Jehovah, who, caring little for a mere Christian, left us to walk the baking street four mortal hours, till just when we were going to buy a tent to sleep in, Lutaif remembered that he had a dear and valued friend who lived in Mogador. Comment upon his memory seemed injudicious. His friend, Mr. Zerbib, a missionary, received us into his hospitable house. We straight forgot our troubles and the Jews' unseasonable feast, and fell discussing with our host whether or not the thing in Europe known as progress had proved a blessing or a curse, sporadically introduced into Morocco by the waifs and strays, the tourists, traders, runaway sailors, and the rest who, on the coast, act as the vanguard of the army of the light.

Chapter II

ALL Mogador we found much exercised about the province of the Sus, the very province in which the inaccessible Tarudant, the city of our dreams, was situated. It seemed that about eighteen months ago, one Abdul Kerim Bey, an Austrian subject, had arrived and hoisted his flag as Patagonian consul. Brazil and Portugal, Andorra, San Marino, Guatemala, Haiti, and San Domingo, Siam, the Sandwich Islands, Kotei, Acheen, the Transvaal, Orange Free State, and almost every place where there was revenue sufficient to buy a flag and issue postage stamps for philatelists, had long ago sent consuls to Mogador. Their flag-staffs reared aloft looked like a mighty canebrake, from the sea; their banners shaded the streets after the fashion of the covering which the Romans drew over their amphitheatres, and half the population were consuls of some semi-bankrupt state. Yet Patagonia, even in Mogador, excited some surprise. Jews who had been in Buenos Ayres (and a considerable quantity emigrate thence from Mogador) argued that Patagonia was under the authority of the Argentine Republic. Those who had been in Valparaiso said it belonged to Chile. Few knew where Patagonia really

was. The Arabs, whose geography is fragmentary, thought
" Batagonia " was situated somewhere in Franguestan, and
that contented them. What struck their fancy most was
certainly the colour and design of the new oriflamme.
Barred white and blue, a rising sun grinning across three
mountain tops, a cap of liberty, and a huanaco ruminant;
an Araucanian Indian in his war paint in one corner, and
here and there stars, daggers, scales, and other democratic
trade-marks, made up a banner the like of which had
seldom been observed in all the much be-bannered town of
Mogador. The owner of this standard and the defender in
Morocco of the lives and liberties of Patagonian subjects,
dressed like a Turk (long single-breasted black frock coat
and fez), and spoke a little Turkish, but no Arabic. His age
was that of all the world, that is, somewhere between
twenty-eight and fifty, and his appearance insignificant, all
but his eyes, which some declared to have the power of
seeing through a brick, and others that of piercing through
cloth and leather, and discerning gold from silver in the
recesses of a purse. Be all that as it may, a travelled man, a
doctor; that is, "tabib," for the two words, though given in
dictionaries as the Arabic and English for the equivalents of
the same thing in either tongue, in point of fact, are different. " Tabib," in Oriental lands, is a convenient travelling
designation, as " Colonel " in the Southern States, and as
" Captain " was in Georgian times; it rather indicates a
status than a profession, and, in addition, is not out of place
upon a traveller's card.

Dr. el Haj Abdul Kerim apparently enjoyed his designation and his "tabibship" by the grace of God. His consulate he held by virtue of a mandate of an extraordinary potentate.

Some two-and-twenty years ago, a Frenchman, one Aurielle de Tounens, a man of family and an advocate of Bordeaux, succeeded in persuading the Indians of Arauco that he was their king. This soon aroused the anxiety both of the Chilians and the Argentines; for from the time when first Ercilla wrote his "Araucana," the Indians of Araucania had been free, and if they had had a king perhaps they would have taught the neighbouring Republics what liberty really meant. For a brief space De Tounens flourished under the style and title of "Aurelio Primero, rey de los Araucanos," and then diplomacy or treachery, or both, ousted him, and he died "i' the spital" in his own town of Bordeaux. During his efflorescence he coined money, designed a flag, gave titles of nobility, and appointed consuls; and it appears that one of them was this Kerim Bey, the Turco-Austrian, who swooped upon Mogador, out-consuling all consuls hitherto known by the size and pattern of his flag. It is not likely that Aurelio Primero ever heard of Mogador, still less that from Arauco he sent a special envoy to such a place. Most probably he sent out consuls generally, after the fashion of bishops *in partibus*, with a roving consulship, and with instructions to set up their flags wherever they found three or four dozen fools assembled and a sufficient roof to bear the pole.

One of these roving consular commissions no doubt was given to Dr. Abdul Kerim Bey, in days gone by, in South America. Indeed, I fancy I remember at Bahia Blanca a forlorn Austrian who was said to have held some illusory employment about the person of the Araucanian King, such as head bottle-washer, holder of the royal stirrup, or guardian of the royal purse, the last, of course, a sinecure which, in all courts that have no money, should be abolished in the interest of economy. Our Mr. Abdul,* during his residence in Mogador, having heard that the province of the Sus was rich in mines, and that no port open for European trade existed south of Mogador, the grandfather of the present Sultan having closed Agadir (formerly known as Santa Cruz when in possession of the Spaniards), bethought himself that it would be a master stroke to make a treaty with the semi-independent chieftains of the Sus, open either Agadir, Asaka, or some other port, and trade direct with Europe. Sus being mainly peopled by Berber tribes, who, it is said, are the descendants of the Numidians, who certainly were in possession of the country at the epoch of the invasion of the Arabs, has always been but ill-affected to the central government. The town of Tarudant and the Zowia of Si Hamed O'Musa have hitherto been the two chief centres of the Shereefian† authority, but recently, from some fancied slight or from ambitious motives, the repre-

* Abdul Kerim means slave of the Merciful, merciful, of course, being an attribute of Allah.
† The Sultan of Morocco is called His Shereefian Majesty, as being a shereef, i.e., descended from the Prophet.

sentative of Si Hamed O'Musa, one Sidi Hascham, has wavered in his allegiance to his lord.

That which is most desired by every Arab intriguer is the possession of good rifles, and it appears that Kerim Bey, Esq., promised to help the chiefs to unlimited supplies of Winchesters. But be this as it may, Kerim appeared in London with a treaty, real or supposititious, a fez, some twenty words of Arabic, several tons of assurance, and the experience of five-and-forty years. With these commodities he got a syndicate together to engage in trade with the province of the Sus, open a harbour, divert the caravans which now come from the interior and the south to Mogador, supply the ingenuous natives with rifles, Bibles, Manchester " sized " cottons, work the real or hypothetic mines, introduce progress—that is electric light, whisky, and all that—and give the acrobats of Taseruelt a reasonable music-hall which might spare them the long voyage to London to seek a fitting place in which to show their powers.

The necessary gentlemen (tribe of Manasseh) with money being found, it was incumbent to get a man de pelo en pecho,* as the Spaniards say, to visit Morocco, see the Sultan, go to the Sus, and arrange matters with the various chieftains personally. Like all the world, Abdul Kerim had many faults, but amongst them the fault of rashness was not numbered. In his wildest moments he had never asserted that he personally had penetrated in the little visited Sus;

* Pelo en pecho—hair on the chest, by inference a brave man, or man of action.

it was thought if the treaty (which he exhibited, but could not read) was genuine, that it had been negotiated by a third person who knew the country well.

Brave men are not so far to seek in London, and one, Major Spilsbury, soon volunteered. He was the kind of man able and willing to walk up to a cannon's mouth; the sort of man who risks his life ten times a day for forty years to gain his livelihood, and dies—either by an Indian arrow, Malay parang, or Arab bullet—"one of our pioneers of empire " or else a " foolish filibuster," according as he succeeds or fails. Quiet and courteous, a linguist, and brave to rashness, he was the very antithesis of Abdul; but such as they were they started out together on their quest.

Sus being the most southern province of Morocco, Abdul Kerim quite naturally went to the north, and dragged his wondering companion all round the empire till at last they found the Sultan, who was in Morocco city, when it turned out that all the boasted influence Abdul Kerim had set forth he possessed was nothing, and the Sultan refused permission to trade direct with Sus. At Mogador the inevitable quarrel took place, and Abdul started for Montenegro, Muscat or elsewhere, and left poor Major Spilsbury alone. He being determined to see the adventure through, engaged a Jewish interpreter, went to the Canaries, chartered a schooner and landed at Asaka after having entered into negotiations from Mogador as to his reception with the chiefs.

Eight or ten days he fought with contrary winds aboard his little unseaworthy schooner, reached Asaka, landed, was well received, made treaties with the chiefs, and all went well until an inferior chief, either being bribed by the Sultan or because he did not think himself sufficiently regarded, or because of the old antipathy to Christians, ever so strong amongst Mohammedans, rode up at the head of fifty horsemen and spread confusion amongst the assembled natives, declaring that he would permit no Christian to traffic in the land. Shots were exchanged, and Spilsbury, bearing his treaty, as Camoens bore his poems, had to escape on board his schooner and for the present leave the enterprise; though, whilst I write, I should not be surprised to learn that he was near Asaka with a fresh expedition.* Quite naturally the advent of such a consul from a new country such as " Batagonia,"† his flag, his fez, his name, and all the subsequent proceedings created some excitement in such a quiet place as Mogador. Consuls assembled, took counsel, wrote dispatches, charged for stamps, and generally fulfilled the functions of their office. Long-bearded Jews, whose talk up to that time had never strayed from money, now dis-

---

* Events proved that I was right, and almost as I was writing news came that Major Spilsbury, in *The Tourmaline*, had tried to land at Asaka again, and had been repulsed by the Sultan's troops, and exchanged shots with the Moorish armed transport, *The Hassam*.

Four of his men were taken prisoners, several of the friendly Arabs were killed, and many others, including the Sheikh Neharek-ou-Ahmed, were sent in chains to linger in the prisons of Fez and Mequinez; Major Spilsbury was detained some time at Gibraltar, and the whole result of the expedition was that the reputation of England was much damaged in Morocco, and the country rendered still more difficult of access than before. I do not hear that those who fitted out the expedition have suffered in any way except by loss of money, but that is, probably, the only kind of loss they could ever feel.

† There being no P in Arabic, the Arabs use B precisely as if they were inhabitants of " Botzen und Bosen."

cussed questions of diplomacy, of international law, discoursed on piracy, of filibustering, wondered if business would have been improved if Spilsbury had got a footing, but thought no affair sublunary could be quite rotten if Daniel Sassoon* had had a finger in the pie.

The Arabs generally were puzzled, but pleased to think there seemed a chance of good repeating rifles being in the market; but all the European residents saw clearly the hand of all-encroaching England, and in especial the French were certain that Mr. Curzon had given his sanction to a plot to extend the dominion of the empire over Sus.

All things considered, it was a most inauspicious time to attempt a journey to such a place as Tarudant, guarded most jealously itself from Europeans both by the fanaticism of the inhabitants and by the special prohibition of the Moorish Government to any European to pass south of the Atlas Mountains to the plains. Hardly had we well landed in the town, when a report was spread that we were agents of the British Government, or advance couriers in the interest of the syndicate. Our sojourn at the house of so well-known a missionary as Mr. Zerbib in some way allayed the public fear, for no one credited him with any but purely spiritual views of conquest. The fact that we had no arms, suggested our great cunning, for no one doubted that we could lay our hand on stacks of rifles if we chose, though how we could have done so was a mystery, as all the influence of our minister at Tangier proved insufficient to procure me even a

---

* Mr. Sassoon was reported to be interested in the venture.

pass for a common double-barrelled gun, so much alarmed were the authorities after the recent landing on the coast. No one, except the Turks, more clearly comprehend that only the jealousy of the European nations saves their independence than do the ruling classes of Morocco. They understand entirely the protestations about better government, progress, morality and all the usual "boniment" which Christian powers address to weaker nations when they contemplate the annexation of their territory. On the one hand they see the missionary striving to undermine their faith, and on the other they behold the whisky seller actually sapping their Mohammedan morality; behind them both they see the ironclad arriving in their harbours under frivolous pretences to exact enormous compensation often for fancied injuries, and they well know the official Christian's God is money, or as they say themselves, "amongst the Nazarenes all is money, nothing but money."* The Moors have vices, plenty of them, some of them well-known in London and in Paris, but in their country a poor Mohammedan, unless in case of famine, is seldom left to starve. Even a begging Christian renegade, of whom there are a few still left, always receives some food where'er he goes, and is not much more miserable than the poor Eastern whom one sees shivering about the docks in London and imploring charity for "Native Klistian" with an adopted whine, and muttered national imprecation on the unsuspecting almsgiver.

* "El N'zrani kulshi flus" is a common saying in Morocco.

The Moors all know when once a European gets a footing in their land, even although that should be brought about by filibustering syndicates financed by London capitalists, that the nation to whom the filibusters belong steps in to guard its subjects, and having once stepped in, remains for ever. They see Ceuta, Alhucemas, el Peñon de la Gomera and Chafarinas all in foreign hands, and like the fact as much as we should like the Russians in the Isle of Wight. Therefore, their irritation about the Sus was most intense, and the jealousy of foreign travellers never keener.

Much has been said about the badness of the Government of Morocco. Most Governments are bad, the best a disagreeable institution which men submit to only because they fear to plunge into the unknown, and therefore bear taxation, armies, navies, gold-laced caps, and all the tawdry rubbish which takes from themselves, to furnish living and employment for their neighbours, under the style and title of national defences, home administration, and the like.

In countries like Morocco, where men still live under the tribal system, all government must be despotic; witness Algeria, Afghanistan, and Russian Tartary. The unit is the tribe and not the individual, and what we understand by freedom and democracy would seem to them the grossest form of tyranny on earth. No doubt no man in all Morocco is secure in the enjoyment of his property; but then in order to be amenable to tyranny one must be rich, and as most tribesmen own but a horse or two, a camel, perhaps a slave,

some little patch of cultivated ground or olive garden, it is
not generally on them the extortion of the Government
descends, but on the chief Sheikh, Kaid, or Governor, who,
if he happens to be rich, can never sleep secure a single day,
for he knows well some time he will be brought to Fez or
to Morocco, thrust in a dungeon underground, and made to
give up all he has on earth. True, whilst this very man
enjoys his wealth and place he oppresses all the tribe to the
utmost of his power; but still I fancy that hardly a Moor
alive would change the desultory Eastern tyranny, which he
has suffered under all his life, and under which his fathers
groaned since the beginning of the world, for the six-
monthly visit of the tax collector as in Algeria. When people
in Morocco speak of Algeria they admit the safety of the
roads, the gathered harvests, no hostile intervention coming
between the sickle and the wheat; they admire the railroads,
laugh at the figure which the French soldiers cut on their
horses, but generally finish by saying, " the Arab pays for
all, and in that land they tax your dog, your horses, and you
yourself, and all are slaves."

Most Europeans point with pride to the curious system known as " protection," and called by the Arabs
" Mohalata," which for at least a hundred years has existed
in Morocco, as something to be proud of. The system needs
a few words, as generally writers on Morocco, without a
word of explanation, talk of the custom and state it is a
good one, in the same way that free and fair traders each
assume their nostrum is the best, or as professors of the

Christian or Mohammedan religions look on their dogmas
as being alone fitted for honest men to hold. Briefly, the
system of the " Mohalata " was invented to obviate the difficulties of trading in a country so badly governed as is
Morocco. The word in Arabic means partnership, but the
system has been complicated by the habit of protection by
means of which the European partner generally contrives
to get his Moorish partner made a citizen of the country to
which he himself belongs. Thus, " Mohalata " and " Protection " have come to be so mixed together, that it is rare to
find a Moor in partnership with any European and not
protected by a European Consul. Once protected, the Moor
ranks as a Montenegrin, Paraguayan, Englishman, Frenchman, or Portuguese, or what not, and is removed from the
exactions of his own Kaid (governor), and even is placed
outside the jurisdiction of his own Sultan.

So far, so good, for no one can pretend the Sultan's government is good, and under shadow of the protection system
many individual Moors have become rich. But in their efforts
to escape from their own rule, the wretched Moors often
fly from the claws of Moorish tribal feudalism, and fall into
the mouth of European commercialism, unrestrained by
public opinion, the press, or any of the preventive checks
which keep the " cash nexus " system within some sort of
bounds in England. The following anecdote may serve to
illustrate how the protection system occasionally works out.*

* I will be glad to give names in full to anyone who will take up the poor
devil's case.

Mohammed —— ten years ago was partner of a European merchant in Mogador. The European (a God-fearing man) purchased three camels on condition that Mohammed —— should act as camel-driver, and take the beasts about the country wherever it was profitable to take goods. A camel in Mogador may be worth some thirty dollars. For this business Mohammed was to receive a certain portion of the profits made. For several years all went well; Mohammed made his journeys, took his merchandise about, and got his portion of the gains, feeding the camels at his own expense.

One day the Christian (in Morocco all Europeans pass colloquially as Christians) said to Mohammed, "I intend to leave the place and to return to Europe. My intention is to sell the camels, and we can then divide the profits of the sale according to our deed." The Arab answered he was willing, and began to cypher up the sum the beasts should bring when sold.

The Christian then informed him that he had a scheme by which he thought that they might each gain much, and if it prospered, that Mohammed could keep the camels for his pains. Mohammed, nothing loath, sat all attention to hear the expected plan by means of which he was to keep his beasts. "You shall take the camels," said the Christian, "and load them for a journey to the Sus. At some point of the journey thieves shall attack you, and you shall then throw all the merchandise upon the ground, then return home at once, and swear before the Cadi that I entrusted you with two thousand dollars and it is stolen, and I will force the Government to compel the Sheikh of the tribe where the robbery was done, to make all good, and we will share the money, and you can keep the camels for your own." A scheme of this kind always attracts an Arab; it is just the sort of thing he would invent himself. And when his own ideas are returned to him, passed through the medium of a Christian mind, he is certain that the thing must be all right. Curiously enough, although the Moors are never tired of cursing at the Christian sons of dogs, yet they are well aware of their superior business capabilities, and never seem to doubt their word in matters of the sort. "But," said Mohammed, "if I tell the Cadi that I had your money and that thieves took it, he will throw me into prison, and there is little chance I shall ever come out alive." "Have no fear," said the merchant, "the imprisonment will be a mere formality. I will feed you when you are in prison, and in a few days you will be free."

The camels were duly loaded, and Mohammed set out upon his journey to the Sus. In a few days he returned, having torn his clothes, rolled himself in the sand, and with some self-inflicted bruises, informed his friend the merchant, who took him to the Cadi to testify on oath.

Most unluckily for the miserable man the place he chose to pitch upon for the scene of his adventure was a few miles outside the town, in a district called Taguaydirt. The Cadi sent for the headman, who came and swore that he had never seen Mohammed, and he himself failed to identify any of the natives of the place, who were presented to him.

Seeing the thing looked grave, he went and took sanctuary in the tomb of Sidi M'Doul,* the patron saint of Mogador, about a mile outside the town. Inside the sanctuary the man was safe, and every day his European friend sent him his food, his " Tajin,"† " Couscousou,"‡ flat Moorish bread, and green tea (called Windrisi from Windres, that is London, from whence it comes), seasoned with mint and sweetened with enormous lumps of sugar broken with a hammer from the loaf. A week passed by, and every day his wife and children came and talked to him, standing outside the shrine, and much elated at the kindness of their European friend, and of the affluence which, in a day or two, was to burst on them through his influence.

But all their feasting did not suit the European's book, and he contrived to get Mohammed out of the sanctuary, upon the pretext that it was necessary to swear again to the affair before the judge. The swearing and examination over, the Cadi (at the Christian's instigation) threw the poor devil into prison, and then for a few days the Christian sent him his food, as he had done before when in the sanctuary. After a day or two he feigned to think Mohammed had deceived him as to the robbery, and had really taken the two thousand dollars for himself. The supply of food then

* Sidi M'Doul is said to have been a Scotch sailor who became a Moor and after his death, a saint. Be this as it may, from the saint's name Europeans have made the name Mogador, which is never used by the Moors, who call the town Sueira, The Picture.

† Tajin literally means "the dish." It is generally a greasy stew of mutton, soaked with rancid butter and saffron, and seasoned with asafœtida.

‡ Couscousou is a kind of dry porridge made of grated wheat, stewed, and served up with mutton or chicken, and pieces of boiled pumpkin.

ceased, and the Christian raised a plea against the Arab for restitution of his money, or, failing that, the seizure of his goods.

The wretched man, seeing himself deceived, confessed the plot, but as he (this time) spoke the truth, no one believed him. The lawsuit ran its course, and the Arab's wife sold off his horse and gun (the most cherished property an Arab has), sold off his camels, their cows, their goats, and sheep, and raised six hundred dollars after selling everything she had. His children begged, the wife worked as she could, the husband, heavily chained in prison, starved.

Five long years passed away, the wife feeding her husband as she could, the children running about like pariah dogs, maintained by charity. Then the good Christian merchant died, and his heirs, of course, still pressed for payment of the debt.

Four more long years went by, and then a thing occurred which makes one think of the proverb that " to jump behind a bush is better than the prayers of good men."* Within the prison were five hundred men, Mohammed one of them; a mutiny broke out, the guard was overpowered, and a few dozen men got out. Then came Mohammed's turn, and he, thinking that his good deed might win his freedom, seized the key and locked the door, keeping the rest within. News went unto our lord the Sultan in his camp, and people hoped that he might exercise his clem-

---

* "Mas vale salto de mata que ruegos de hombres buenos," goes the adage in Spanish, and it is one that most sensible men will endorse.

ency. Back came a letter praising Mohammed's deed, and saying he deserved his liberty, but that the Sultan could not grant it till the debt was paid.

Ten years have passed away, the merchant is long dead, six hundred dollars paid for nothing, a family reduced to misery, and still the victim of the plot remains in prison, heavily chained and prematurely old; Allah looks down, the call to prayers rises to heaven five times a day, and poor Mohammed, a grey-headed man, resigned and uncomplaining, talks to the casual stranger at the prison gate and says the Christian was no doubt a knave, but that the thing was written (mektub), and that therefore no one was to blame, Allah Ackbar, no God but God, and Lord Mohammed is his messenger.

The protection system may benefit the Jews, who, once despised and spit upon by every Moor, have of late years become the tyrants of the land. Scarcely a Jew of any property in any Mellah* in Morocco, who is not a citizen of some foreign state. Perhaps America has made the most use of its protective powers. Both the United States and the Brazils have frequently named consuls who were quite unworthy of representing either state. These men, in several instances, have sold protection right and left, and nothing is more common than to meet a Moor or a Jew in one or other of the seaport towns, who tells you that he is a Brazilian or

---

* Mellah is the word used to designate a Jewry in Morocco. Literally it means Salt, and I have never heard any explanation of the term, but the salt has not lost its savour, as any traveller not suffering from rhinitis can testify.

an American. Today the United States seem to have seen the error of their ways, and several of their consuls are of high character, and fitted to do honour to the post they occupy.

Until quite recently, at times a consul would "sell a Moor," that is would tell his luckless "citizen" that, unless by such a time a sum of money was forthcoming his protection would be withdrawn. The effect of this would be·as a sentence of death to the unlucky man. Generally the protected citizens amass some money, and if the protection is withdrawn, their Kaid or Governor falls down upon them, puts them in prison, where they either die or else remain till, as the prisoner in the Gospel, they have paid the uttermost farthing they possess. As to reclamations, how can a Moor, speaking no foreign language, go to Andorra, Montenegro, or San Marino, to appeal. Pay and appeal,* the proverb says; but fancy an Arabic-speaking man, without a cent, appealing in New York, or in Brazil, in neither of which countries men of dark races are viewed with favour, and justice is a costly pastime even for the rich.

No doubt some few have become rich under protection. Witness the case of Si Bu Bekr, who for so long a time was British agent, and who, when a few months ago he showed me all his treasures in his palace in Morocco city, tapped on his iron chest, and said, "This one is gold, that is all jewels; this, again, is full of bonds"; and is assumed

* Pagar y apelar.

to have a hundred thousand pounds all safely tied up in Consols.

But, on the other hand, sometimes the partnership and protection is as a shirt of Nessus, and I have heard an Arab say, " Can I not get away from his cursed ' Mohalata ' ? Rather the exactions of my Kaid than the insidious bleeding by my Christian partner." Still it must be confessed that both protection and " Mohalata " are much sought after by the natives, and nothing is commoner than to be asked, whilst on a journey, either to protect or enter into " Mohalata " with a Moor. As in the case of the kindred system which prevails in Turkey, of " capitulation," much abuse creeps in, and as the country is not ripe for mixed tribunals I suppose the chicken will have to live and bear its pip.*

Peacemakers and reformers pass a thankless life, and it appears—as almost every ill we see is irremediable, and as the world goes on quite cheerfully (no matter what we do), crushing the weak, and opening wide upon the passage of the strong—that curses are no use; the only course the wise man can adopt is to stand well away and keep his own opinions to himself, unless, indeed, after the fashion of the man in Joseph Conrad's story,* he prefers to hang himself, and then put out his tongue at the world's managing directors.

* " Viva la gallina con su pepita."
† " Outpost of Progress," Cosmopolis, June, 1897. Story of an outpost of Progress told without heroics and without spread-eagleism, and true to life; therefore unpopular, if indeed, like most other artistic things it has not passed like a " ship in the night."

Finding myself the observed of all observers in Mogador, I transferred my residence to Mr. Pepe Ratto's International Sanatorium, about three miles outside the town, which passes generally under the designation of the Palm-Tree House. There I essayed to live my filibustering character down, and for a day or two went sedulously out shooting in the hottest time of day, to show I was a European traveller; collected " specimens," as butterflies and useless stones; took photographs, all of which turned out badly; classified flowers according to a system of my own; took lessons in Arabic, and learned to ride upon the Moorish saddle. A few days of this exhilarating life made all things quiet, and the good citizens of Mogador were certain that I was a *bona-fide* traveller and had no design to attack the province of the Sus.

The Sanatori Internacional de la Palmera was a sort of hotel of the next century. Everything in it was " *en construction*." The managers, two little Marseillais, of the bulldog type, spent almost all their time either in practising *la boxe Marseillaise,* in playing on the concertina, an instrument which, when I am in Europe (dans les pays policés), I fancy obsolete, but which, in days gone by, set my teeth often aching in the River Plate and in Brazil. After so many years when first again I heard its wheezy tones, upon a moonlight night in the Palmera, with camels resting under the great palm tree, and Arabs lying asleep, their faces covered in their haiks, horses and mules champing their corn, hyenas growling in the distance, jackals yelping, and the frogs croaking like silver cymbals, as they never croak to the north of latitude forty, it set me wondering why men must go about on a calm, clear night grinding an instrument to make their unoffending fellows' stomachs ache. Besides the concertina and " la boxe " (Marseillaise), the brothers, curly-headed and pleasant little sons of La Joliette or La Cannebière, devoutly entered everything into a ledger, large enough for Lombard Street, by double entry; and besides that had an infinity of talents *de société*, kept chameleons, understood botany, were cooks and linguists, speaking most languages including " petit nègre " quite fluently; were civil, educated, ignorant, and thoroughly good fellows to the full length of their respective five feet four and five feet seven inches.

The hotel was on a hill and had a view over a sand hill, on which grew oceans of white broom, dwarf rhododendrons, gum cistus, thyme (which in Morocco is a bush), and mignonette, and in whose thickets wild boars harboured and from which sand grouse flew whirring out. The owner of the place, a mighty sportsman, having slain more boars, and had more adventures in the slaying than any one, since Sir John Drummond Hay laid down his spear. Born in Mogador, of English or Gibraltarian parents, and speaking Spanish, English, Arabic, and Shillah quite without prejudice of one another, Mr. Ratto, known to his friends as " Pepe," fills, in South Morocco, the place that Bibi Carleton fills in the north. No book upon Morocco is complete without a reference to both of them.

How the thing comes about I do not know, but not un-
frequently the sons of Europeans born in hot countries
turn out failures, either in person or in mind, or both, but
when the contrary occurs and the transplanting turns out
well, the type is finer than is common in the mother
country. Both of my types would, walking in a crowd in
any town of Europe, attract attention. Tall, dark, brown-
eyed, erect and lithe, clear brown complexions, open-
handed and quick of apprehension, good horsemen, linguists, and yet perhaps not fitted to excel in England or
in France, or any country where continuous work is necessary, they have had the sense to stay at home, and become
as it were " Gauchos," that is a sort of intermediate link
between the Arab and the European, and at the same time
to incorporate most of the virtues of the two races. Put
them in Western Texas, Buenos Ayres, or South Africa,
and they must have made fortunes; as it is both are as
rich as kings when mounted on a good horse, a rifle
in their hands, and a long road to travel for no special
cause.

Not far away begins, sporadically, the district of the
Argan Tree,* in fact, outside the door of the Palmera

---

* The Argan Tree, the Elœdendron Argan of some and the Argania Sideroxylon
of other commentators, for botanists like doctors often disagree, seems to belong
to the family of the Sapotaceæ. Its habitat is very limited, being apparently confined to the sandy district between Mogador and Saffi, in which it forms a dense
wood stretching for forty miles. In habit it resembles an Acacia, being thorny,
twisted in trunk and limbs, and able to survive the longest droughts without
apparent suffering. It produces a nut about the size of an olive, from which an oil
is extracted and used in cooking by the Moors. It is unpalatable to those Europeans
who have never eaten a Turkey Buzzard.

stands a small specimen, the roots almost uncovered and bent towards the east by the prevailing wind.

Not far away, still lives a patriarch of this restricted family, flat topped and gnarled and like a Baobab, its branches taking root all round the stem, and running on the ground for fully fifty feet, goats climbing on its limbs, snails clinging to the leaves, pieces of rag tied to the boughs by passing Arabs, reminding one of the Gualichu* tree in the South Pampa of Buenos Ayres. After long years of life it seems to rest, putting forth leaves and shoots, and bearing fruit, as if it were by habit, and as a protest against the decay which has overtaken all its fellows. The passing Arab, though he may break a branch to light his fire, still reverences it in a vague way, never forgetting as he passes to praise God for it, as if Allah had set it there to tell him of his power.

The choice of guides became a difficulty. Few men in Mogador cared to attempt a journey to Tarudant in company with a European, even though disguised. Arabs who knew the way were terrified at venturing alone into the territory of Berbers, and Arabs feared to be found out upon the road and put in prison by the Sultan's governors.

All were agreed the journey was hazardous, although

* Gualichu is the god (or devil) of the Pampa Indians. At any rate, he is the spirit they propitiate by tying rags, cigars, pieces of hide, tin cups—or anything they may have useless enough to be offered to a diety—to its branches. The tree, which I take to be a Chañar (Gurliaca Decorticans), though others, perhaps wiser than myself, say it is a Tala, stands on a little eminence, and is the only tree for leagues. Darwin remarked it and camped close by it, and it is known all over the South Pampa from Tandil to Patagones.

to what extent they were not sure. Sometimes in travelling in Arab countries it is possible to take a guide into a certain part, and then to get some tribesmen to accompany you. In our case this was impossible, as I could not speak Arabic sufficiently to pass off as a native of the place. Even to say I was a Georgian, a Circassian, or Bosnian, for any of which I might have passed as far as type goes, would have aroused suspicion, for why should an inhabitant of such a country journey to Tarudant?

Although the place I wished to visit is tabooed for Europeans, still Arabs, like other men, delight in doing what they know they should not do, with the full consciousness of doing wrong.

To the illiterate Moor or Arab nothing seems so wrong as to eat bacon, pork, or touch a pig, and yet at times they say "I ate some pork the other day, it was magnificent," after the fashion that boys smoke at schools, pretend to like it, and are sick behind a hedge. An Arab says no wonder Christians are so red in the face and look so well, do they not feed on pork and drink strong wine? It seems to be implanted in the human mind that anything a man is bidden not to do, must, of necessity, be the one thing that if he did it, would make him happy all his life. If this be so, and clergymen (all of the highest moral standing) have assured me that is the case, surely the general concensus of the opinion of mankind is towards doing everything they like, and if that is the case it must be right, for anything which can secure a majority of votes is sent from heaven, for God himself is quite uncertain of the justice of his acts till men have voted on them.

Still, guides for such a journey did not abound. One was too old, a second too religious to go with Christians, or a third too big a rogue even to go with Christians; till at the last a man, Mohammed el Hosein, came forward of his own accord.

Report averred he was a slave-dealer, but the best muleteer in the country. In person he was thin and muscular, age thirty-eight, just married, a first-rate horseman, cunning and greedy, but to be depended on if once he gave his word. All his delight, as he himself informed me, was to drink green tea and smoke tobacco, and, therefore, like the old Scotch lady, who, when a cook was recommended to her for her good moral character, exclaimed, " Oh, damn her morals, can she mak guid broth?" I did not boggle at his slave-dealing, but took him on the spot. Strangely enough he had been employed by missionaries, who spoke well of his capacity touching his muleteership, but lamented over the hardness of his heart. By nationality he was a Berber, with the thin face, small eyes, and high cheekbones of all his race. He sang in Shillah, in a falsetto voice, a quavering air, both in and out of season, and seated on a mule, a packing needle in his hand to act as spur, got over more mortal leagues of country in a day than any other mule driver whom I remember this side of Mexico.

After the muleteer, came choice of roads, for three were

open to us; the first along the coast passing the town of Agadir.* This road is flat and sandy, and follows close to the coast right down to Agadir, and by it Tarudant can easily be reached within five days from Mogador. The disadvantages of following this road were three; firstly, we had to pass the town of Agadir, in which the Sultan had a governor; and secondly, Agadir once passed, we had to traverse the country of the Howara tribe, which bears an evil name for turbulence. Journeying, as I proposed to do, in Moorish dress, two difficulties lay in my path.

Firstly, I might be recognized, and if so recognized by an official of the central government I should have been turned back at once, as has already happened to other travellers in South Morocco. Again, dressed as a Moor and not discovered, I had to run all the risks a Moor must run in travelling, from robbers and from violence. These risks do not beset a European travelling in European dress to the same degree, as Moors in general are chary of meddling with Europeans. It may be asked, why then did I adopt the Moorish dress, and not go boldly as a European after

* Agadir Ighir (Ighir means a fort in Shillah) was once held by the Spaniards, and called Santa Cruz. It is situated on a slight eminence near Cape Gher, has a tolerable port and is the natural outlet for the trade of the Sus, but it is closed to trade by order of the Sultan, and the merchants in Mogador do all they can to keep it closed, as they themselves depend much on trade with the same province. In the last century Agadir had a flourishing trade with Europe, but the closing of the port killed the place, and there are now not above a thousand inhabitants. Amongst these there are a good many Jews, and it is reported that amongst these Jewish families there are to be found the handsomest women in Morocco. One regrets that there is no trade with Europe, on account of these daughters of Israel.

having got a permit from the Sultan, and taken a guard of Moorish soldiers.

These were my reasons: the Sultan of Morocco, when he gives a European a permit to travel in his territory always writes on it, after the usual salutations to his various governors, "We recommend this Christian* to you, see that he runs no danger." The Moor, reading between the lines, sees that the Sultan wishes him to stop the Christian visiting any unfrequented place, and naturally he puts a lion in the path. Thus, had I gone to Agadir furnished with guard and Sultan's letter, the governor would have received the letter, kissed it, duly placed it on his forehead, called to his scribe to read it, made me welcome, and informed me that it was quite impossible for me to go farther, as certain bastards,† who feared neither God nor Sultan, would be sure to kill me on the road. I should have told him: "Well, my death be on my own head," and he would straight have answered, "Be it so in God's name, if it were only yours, but who will shelter me from the anger of our Lord the Sultan if you are killed?" Persuasion, bribes, and everything would have been in vain, and had I then insisted, I should have found myself politely escorted back by a guard of cavalry, as all the Sultan's governors are well aware that their liege lord admits of no mistakes, but punishes mistakes and faults alike.

* Rumi, Roman, is the polite word for a Christian. N'zrani or Nazarene is half-contemptuous.
† Oulad el Haram.

Just as I had determined to risk the journey by the way of Agadir, news came that the Howara tribe was in rebellion, and that the road was shut. There still remained two mountain passes through the Atlas, one starting from a place called Imintanout, and going by Bibouan to Tarudant, and still another from Amsmiz, a town close to Morocco city, which crossed the Atlas at its greatest breadth and led to Tarudant, across the River Sus at a place called Ras el Wad, a whole day's journey from Tarudant. Needless to say both roads were longer and much more difficult than the coast route, and by either of them I should have employed at least eight days to reach my destination.

Choosing the shortest road I then determined to go by Imintanout, and set about at once to make my preparations for a start.

Mohammed el Hosein brought his own mule and with it another belonging to one Ali of the Ha-Ha* tribe, that is to say Mohammed hired a mule from Ali who accompanied us (as I learned upon the road) to see his animal was not ill-treated. Ali I suspect received no pay, but was a sort of general *homme de peine,* and quite contented so that he received his food and that his mule was fed, and even thought himself quite fortunate when he received a pair of cast-off shoes. He had no idea where we were going to, and when we told him, wished to return, and would have done so had we not laid hands upon his mule,

---

* Ha-Ha is the name of the province in which Mogador is situated; it is also the name of the tribe.

which seeing, he resolved to brave all dangers rather than trust it to the tender mercies of Mohammed el Hosein. For the rest Ali was a muleteer, which race of men, whether Spaniards, Mexicans, Turks, or Brazilians, is all alike, singing all day while sitting sideways on their beasts, smoking continually, eating when there is food, and sleeping quite contentedly (as the unjust all sleep), their heads resting upon a pack saddle, feet to the fire, and with a tattered blanket covering their faces from the dew at night. Lutaif, following his character of a man of letters, rode on a mule, perched on a high red pack saddle, which, loosely girthed, after the Moorish fashion, swayed about and made it quite impossible for him to mount or to dismount without assistance. By this time we had bought our Moorish clothes, in which Lutaif, being a Syrian, looked exactly like one of the figures on the outside of a missionary journal, which assume to represent biblical characters, and really are a libel on the Syrian race. Having arranged to represent a Turkish doctor, I put on the clothes with some misgivings, and left my room in the Palmera with the air of one who has assumed a fancy dress. On my appearance all declared that I need never say I was a Turkish doctor, for I looked so like a man from Fez, in type and colouring, that it was better to say nothing as to who I was, and that the passers-by would take me for a travelling Sherif. A Sherif being a holy character generally rides upon a horse, and so I purchased one through the good offices of Mohammed el Hosein, who happened to know of " the best horse in all the province which his owner wanted to dispose of—with saddle and bridle all complete—for a mere nothing, being short of cash." I purchased the whole " outfit " at nine-and-twenty dollars, a third more than a native would have paid. The saddle was rotten, and felt like riding on a bag of stones, but the horse, though lean in condition, was full of quite unsuspected spirit, sure-footed, and excellent upon the road. Equipped with horse, high Moorish saddle covered with red cloth, dressed in white, but with a blue cloth cloak to cover all, a fez and turban, head duly shaved, and yellow slippers, with, of course, a pair of horseman's boots (called temag by the Arabs) buttoned up the back with green silk buttons, embroidered down the sides with silk and silver thread, a leather bag to sling across the shoulders and act as pocket, I was ready for the start.

Tents and the general camp equipment of a European journeying in Morocco, did not trouble me. We had a little tent packed on a mule, just large enough to serve as sleeping quarters for myself and for Lutaif. The men had to sleep by their mules after the Moorish fashion, and if it rained to come for shelter under the lea side of our tent. Cooking utensils were but a kettle and an iron pot; we had no forks or spoons, as being dressed as Moors we had to eat after the Moorish fashion with our hands, our only luxuries being a rather gim-crack brass tea tray, a pewter German teapot, and six small glasses to drink green tea flavoured with mint, and made as sweet as syrup. In my anxiety to be quite the native, I even left my camera, an omission which I regret, as had I taken it I might have published a book of views of the Atlas, and saved much trouble to the public and myself.

A European who came to see us off looked at our modest equipage with some disgust, and said that he had never seen a Christian start like a Susi trader, and that we should soon repent the want of European comforts. Your European comfort when in Europe is in place; but on the march in a wild country everything additional you take is as the grasshopper in the adage, which I think soon made its presence felt. It is customary for fools and serious men, when setting out on any journey (say to Margate), finishing books, entering into the more or less holy state of matrimony, becoming bankrupt, or entering holy orders, going to sea, meeting their first love, burying their most disagreeable relation, or being jilted, thrown from a bicycle, being kicked or knighted, in fact in any of the disagreeables, which like rain fall on the just and the unjust, but always show a preference for the poor honest man, to sit down and record exactly how they felt, what thoughts occurred to them, and generally to disport themselves as if another mortal in the world cared they were even born, except their mothers and themselves.

Time-honoured institution, from which no scribbling traveller should depart. If you have naught to say, why write it down, extend it, examine into it, and write and write till after writing you persuade yourself you have written something; and if upon the other hand you happen to have aught worth writing, keep it to yourself, and go to bed remembering that tomorrow is another day, that thoughts will keep and mules are ever better saddled about an hour before the sky on the horizon begins to lighten, and the first faint flush of dawn spreads slowly like a diurnal Aurora Borealis and drives the morning star back to its night; for then, as mules are cold and empty, they cannot swell their stomachs out so much and stop the muleteers from drawing home the girths.

Chapter III

LEAVING the International Sanatorium of the Palmera at the hour that Allah willed it, which happened to be about eight in the morning of the 12th October, dressed in Moorish clothes, our faces far too white, and our ample robes like driven snow, the low thick scrub or argan, dwarf rhododendrons, and thorny sandarac and " suddra "* bushes after five minutes' riding swallowed us up, quite as effectually as might have done a forest of tall trees. Mohammed el Hosein, fully aware of the importance of getting accustomed to the Moorish clothes before at once emerging on the beaten track which leads from Mogador to Morocco city, engaged us in a labyrinth of cattle tracks, winding in and out for full two hours through stones and bushes, following the beds of water courses, dry with the twelve months' drought, which had caused almost a state of famine, and calling to us to hold ourselves more seemly, not to let our " selhams "† hang too low,

---

* Suddra is the Zizyphus Lotus of botanists. It is extremely thorny, and is much used by the Arabs to make the enclosures known as " Zerebas " round their houses; when dry it takes a curious grey-blue tinge, very effective in certain lights. It is of this plant, I think, that the Soudanese make the temporary " Zerebas " round their camps, which on occasions have given so much trouble to our troops.

† Selham is the hooded cloak worn as an outer garment; it is made of blue cloth or white wool. It is the " burnouse " of Algeria.

not to talk English, and when dismounted to walk as befits Arab gentlemen, to whom time is a drug.

After much threading through the tortuous paths, getting well torn and sunburned by the fierce sun, we emerged at the crossing of the river El Ghoreb, which runs into the sea at Mogador.

Here we encountered the usual stream of travellers always to be met with at the crossing of a river in countries like Morocco; grave men on mules going to do nothing gravely, as if the business of the world depended on their doing it with due precision. Long trains of mules laden with cotton goods going up to the capital; a travelling Arab family, the father on his horse, his gun cased in a red cloth case, balanced across the pommel of his saddle; his wife, either on foot or seated on a donkey following him, the children trotting behind, a ragged boy or two drawing a few brown goats, a scraggy camel packed with the household goods on one side, and on the other with a pannier from which a foal stuck out its head; and, lastly, two or three grown-up girls, who, as we came upon them crossing the stream, lift up their single garment and veil their mouths according to the laws of Arab decency. We sit and eat under a tree as far away as possible from all the passers-by, and our clean clothes and look of most intense respectability, secure us from all danger of intrusion on our privacy.

No sooner seated than Swani seized my legs and pulled them violently, and rubbed the knee-joints after the fashion of a shampooer in a Turkish bath. On my enquiry he assured me he knew I must be suffering agony from the short Moorish stirrups and cramped seat. I had indeed felt for the past half-hour as if upon the rack; but a horseman's pride and acquaintanceship with many forms of saddles had kept me silent. The rubbing and pulling afforded intense relief, and I acknowledged what I had endured, on his assurance that no one escapes the pain, and that the most experienced riders in the land are sometimes kept awake all night, after a long day's march, owing to the stiffness of their legs.

In mediæval Spain, good riders were often designated as " Ginete en ambas sillas,"* that is accustomed to either saddle, *i.e.* the Moorish and the Christian, and I now understood why chroniclers have taken the trouble to record the fact. Strangely enough the high-peaked and short-stirruped saddle does not cross the Nile, the Arabs of Arabia riding rather flat saddles with an ordinary length of leg. The Arab saddle of Morocco, in itself, is perhaps the worst that man has yet designed, but curiously enough from it was made the Mexican saddle, perhaps the most useful for all kinds of horses and of countries that the world has seen.

The Moors girth loosely and keep their saddle in its place by a broad breast plate; so that it becomes extremely

---

* This phrase often occurs in Spanish chronicles, after a long description of a man's virtues, his charity, love of the Church, and kindness to the poor, and it is apparently inserted as at least as important a statement as any of the others. In point of fact, chronicles being written for posterity, it is the most important.

difficult to mount, and to do so gracefully, you have to seize the cantle and the pommel at the same time, and get as gingerly into your seat as possible. Like all natural horsemen, the Moors mount in one motion, and bend their knees in mounting; thus, in their loose clothes, they appear to sink into the saddle without an effort. Once in the saddle a man of any pretension to respectability has his clothes arranged for him by a retainer, as being so voluminous, it is quite an art to make them sit.

Swathed in the various cloaks and wrappings which constitute the Arab dress, the feet driven well home into the stirrups, and gripping your horse's sides with all the leg, the seat is firm, though most uncomfortable at first. After a little it becomes more tolerable, but few men can walk a step without enduring agony when they dismount after three or four hours on horseback, especially as it is a superstition amongst the Moors to mount and dismount as seldom as they can, for they imagine the act of getting on and off fatigues the horse far more than the mere carrying the burden on his back. Of course, both getting on and off are done in the name of God, that is after the repetition of the sacramental word Bismillah, used on eating, drinking, riding or performing any action for which a true believer should give thanks to Him who giveth benefits to man. It is the fashion amongst Europeans to sneer at Arab riding, and no doubt an Arab in the hunting field would not look well; and it is possible that a hunting man might also find himself embarrassed to ride a Moorish horse in Moorish saddle fast downhill over a country strewn with boulders, or at the " powder play," to stand upon his saddle and perform the feats the Moors perform.

Horsemen and theologians are both intolerant. Believe my faith, and ride my horse after my fashion, for no Non-conformist, Cossack, Anglican, Gaucho, Roman Catholic, or Mexican can see the least redeeming point about his fellow's creed, his saddle, horse, ox, ass, or any other thing belonging to him.

Lunch despatched, green tea drunk, and cigarette carefully smoked behind a bush, for men in our position must not give offence by "drinking the shameful "* in the face of true believers, we mount again, and plunge into an angle of the Argan forest, which extends from Mogador to Saffi.

Goats climb upon the trees, and camels here and there browse on the shoots; under the trees grow a few Aras (*Callitris Quadrivalvis*), and in the sandy soil some liliaceous plants gleam like stars in the expanse of heaven. After an hour the trees grow sparser and we emerge into a rolling country, and pass a granary, which marks the boundary between the provinces of Ha-Ha and Shiadma, and take our last look of the sea.

The Argan trees become more scarce as we cross into the fertile and well-cultivated province of Shiadma. Sand now gives place to rich red earth, and Swani, pricking his

* " Drinking the shameful " is smoking tobacco, not drinking new whisky as in some civilized lands.

mule with his new dagger, which he had wheedled me
to buy him under pretence that it did not befit my follower
to go unarmed, comes up and asks if, even in England,
there is a better cultivated land. I answer, diplomatically,
that there is none, although perhaps the soil of England in
certain parts is just as rich. Being an Arab he does not
believe me for a moment, but ejaculates, with perfect manners, " God is Great, to him the praise for fertile lands,
whether in England or Morocco."

The Kaid's house on a hill stands as the outward visible
sign of law and order, and Mohammed el Hosein imparts
the information that the prison is always full. 'Tis pleasant
to go back, not in imagination, but reality, to the piping
times when prisons were always full,* maidens sat spin-
ning (I think) in bowers, and the gallows-tree was never
long without its " knot." This leads me to consider whether,
if all the world were regulated by a duly elected county
council, all chosen from a properly qualified and democratic, well-educated, pious electorate, and all men went
about minding each other's business—with fornication, cov-
etousness, evil concupiscence, adultery, and murder quite
unknown, and only slander and a little cheating left to
give a zest to life—they would be happier upon the whole
than are the unregenerate Moors, who lie and steal, fight,
fornicate, and generally behave themselves as if blood circulated in their veins and not sour whey? Despite the

---

*I would not be taken here as wishing to disparage the prisons of my own
country, or to insinuate that they are often empty.

Sultan's tyranny, with every form of evil government thrown in, with murder rampant, vices that we call hideous (but which some practise on the sly) common to everyone, the faces of the poor heathen Moors, whom we bombard with missionaries, are never so degraded as the types which haunt the streets of manufacturing towns. And if the face is the best index to the mind, it may be that the degraded heathen Moor is at the heart not greatly worse than his baptized and educated rag-clad English brother in the Lord.

As evening falls we pass a shepherd close to the high road, sitting down to pray, beside him are his shoes and crook, and not far off his dog looks on half cynically, and up above, Allah preserves his attitude of "non mi ricordo," which is excusable where men worry him five or six times a day. Still the shepherd must have been genuine, and could not have known that infidels would pass his way.

The country here is chiefly composed of red argillaceous earth, the rock limestone, and the general configuration round-topped hills rising towards the Atlas. The Argan trees become more rare, and within sight of our destination we see the last of them.

The Argan, like the Cacti of the Rio Gila, in Arizona, seems to be able to resist any drought. Strange that all-wise Providence failed to endow Africa with either the Cactus or the Aloe, both plants so eminently suited to its climate. It was, however, left to poor, weak, erring human reason to supply the want.

It is pleasing to reflect that for once the powers* generally opposed to one another should have united in endowing a country with two non-indigenous plants, which have taken to the soil as if they had been originally found there. Is it reason after all that is infallible?

Meskala reminded me curiously enough of an "hacienda " in Mexico, with an almost similar name Amascala, the same white walls, the same two towers of unequal height over the gateway, almost the same corrals for animals outside, formed in both cases of the branches of a prickly shrub, goats feeding by the same turbid stream flowing through a muddy channel, and the gate once opened, which in our case took at least a quarter of an hour's entreaty mixed with objurgations, the self-same twisting passage of about twenty yards in length, through which the stranger enters before arriving at the great interior court.

The court, about two hundred feet across, was full of animals, belonging some to the Sheikh himself, and others to the various travellers who had sought shelter for the night within his walls.

We had a letter from the consul in Mogador setting forth that we are friends of his, but not descending to particulars, so that we were ushered into an airy upper room, and bread and butter, in a tolerably lordly dish, was set before us. We were uncertain whether Sheikh el Abbas penetrated my disguise, but if he did he made no sign; nor did it

---

* Reason, I fancy, filtered into man's composition after the original plan was completed, and was maybe the work of the serpent.

matter much, as I intended to be taken for a Christian
travelling in Moorish dress to escape observation (as is
often done), till near the place where we break off into
the wilds, and leave the main road to Morocco city. Had
all gone well, I hoped this would confuse the hypothetic
persons, and jumble up their substance to such an Atha-
nasian extent as to make recognition quite impossible.

Lutaif discourses much of eastern lands and reads el Faredi, an Arab poet, to the admiration of the assembled elders.

Swani makes tea and Sheikh el Abbas drinks the three
cups prescribed by usage, lapping them like a dog, and
drawing in his breath like a tired horse at water, to show
his great content. The upper room looked out upon the
court; and in the moonlight I saw a shepherd, assisted by
a little ragged boy, engaged in separating the goats from
amongst the sheep, and ranging them in two little flocks,
after the fashion that the good are to be divided from
amongst the wicked, when this foolish affair of life is
finished with; though with this difference, that whereas
in this case the two flocks were nearly equal, who can
suppose but that after the last count, the goats will not
exceed the sheep by at least ten to one. In a corral hard
by the horses eat, some camel drivers crouch round a fire,
and as I look at the unchanging Eastern life the call to
prayers reminds me that Allah has blessed it by continuity
for a thousand years.

The Sheikh sat long, talking of things and others, of
the decline of British prestige, the advance of Russia, the

new birth of Turkey, and of the glory of the Moorish kingdoms in the Andalos (Spain); and then of business, and how the Brus (Germans), a nation which he says seemed to have come into the world but recently, from some high mountains, are bidding fair to be the first of Franguestan. The German Emperor strikes him as being a great king. He is a Sultan, says the Sheikh, after the fashion that the Spaniards used to say when Ferdinand VII. had perpetrated some atrocity, " es mucho rey," that is, he is indeed a king.*

I fancy that he knows I am a Nazarene, although my conversation is quite evangelical, that is, yea and nay, and now and then a pious sentence muttered very low to hide the accent. Lutaif and Swani answer for me, as if I were an idiot, and step in, so to speak, between me and the Sheikh, as when he asks, for instance, if I have seen the war ships of the Christians, when they at once respond I have, and give particulars invented at the moment, and I learn that ships steam more than fifty miles an hour, guns carry twenty miles, to all of which I nod a grave assent, and the Sheikh sips his tea and praises God for all his mighty works.

Lutaif tells of a vessel at Beirout, a Turkish war ship, sent by the German Emperor to the Sultan (he of Brus), so large that two young Syrians of his acquaintance, who

---

* Others of his subjects admired his procedure so much that their catchword was reported to have been " Viva Fernando y vamos robando," which after all, is but a practical application of the old Spanish aphorism " Viva el rey daca la capa."

had shipped as sailors and got separated, vainly sought each other for seven years, at night climbing to the masthead, by day descending to the hold, but all in vain, because the vessel was so huge; the Sultan could step aboard of her out of his palace on the Bosphorus and after walking all day land in whatever country he desired. This meets with great approval, and I have to confirm it to the letter, and do so with a nod.

The night is hot and the mosquito hums in his thousand; but the Sheikh as he goes warns us to bar the door, because, he says, "Sleep is Death's brother," meaning that when one sleeps death may be near and yet the sleeper be unconscious of it.

The muleteers retire to sleep beside their mules. Swani wraps up my feet in the hood of one " djellaba," and draws another up to my head ready to cover it when I feel sleepy, and as we lie upon the floor, on sheepskins, watching the moon shine through the glassless window, Lutaif puts out the flickering wick, burning sustained by argan oil in a brass bowl, exclaiming, as he did so, "Oh, Allah! extinguish not thy blessing as I put out this light." How much there is in names; fancy a deity, accustomed to be prayed to as Allah by Arabs, suddenly addressed by an Armenian as Es Stuatz, it would be almost pitiable enough to make him turn an Atheist upon himself. I feel convinced a rose by any other name would not smell sweet; and the word Allah is responsible for much of the reverence and the faith of those who worship him.

We left Meskala early and in rain, which soon was over, and entering on a little bit of desert country, the Atlas range appeared like a great wall of limestone capped with white in the far distance.

For three hot hours we passed through a miniature Sahara, rocky and desolate, stones, stones, and still more stones and sand, a colocynth or two lying amongst the rocks, some orange-headed thistles, Zizyphus Lotus here and there, some sandarac bushes now and then; the horses stumbling on the stones, mules groaning in the sand, and no great rock in all the thirsty land to shelter under from the sun. Three hours that seemed like six, until a line of green appears, a fringe of oleanders on the margin of a muddy stream in which swim tortoises, and by which we lie and lap like dogs, and understand wherefore the Psalmist so insisted on his green pastures wherein his Allah made him lie.

In England your green pastures have no significance, and call to mind but recollections of fat cattle and sheep with backs as square as boxes, in the lush grass between the hedges, as the express whirls past and the stertorous first-class passengers hold up their wine glasses against the light and praise the landscape as they eat their lunch.*

---

* It is true that Herrick saw a certain beauty in our " Meadows," or he would not have written the following stanza, but in his days there were no patent manures:

> Ye have been fresh and green,
> Ye have been filled with flowers;
> And ye the walks have been
> Where maids have spent their hours.

But in Morocco and Arabia green grass means life, relief from thirst, and still today their poets stuff their verses full of allusions to the pastures rare to them, but which with us make one at times long for a bit of brown to break the sea of dull metallic green. Fig trees and olives, oleanders* with pomegranates, and a few palms make an oasis in the little desert, and on a sheepskin spread on some cobble-stones close by a rock, exactly like the one that Moses is depicted striking in old-fashioned Sunday books, the water rushing out in a clear stream, we lie and smoke and fall a-talking of our chances of reaching Tarudant.

Mohammed el Hosein gave it as his opinion that if he could conduct a Rumi† there, he would make his name in Mogador as the best muleteer in all the south, and all his previous fears seemed to vanish as he descanted on the line of conduct to pursue when once inside. He seemed to think the risk, if known to be a Christian, was considerable, and counselled that we should encamp outside the gates and reach the town a little after dawn when people were arriving to sell provisions, and then go instantly to the Governor's house which was close to the gates, and claim protection from him. Swani, who, as a native of Tangiers,

---

* Difla in Arabic, from which the Spaniards have taken their word Adelfa, as from Dib, jackal, they have formed Adibe, Berk, a pond, Alberca and the like.

† I.e., Roman is generally used by Arabs in North Africa when they wish to be civil to a Christian. " Caballer el rumi " has a pleasant sound, even when uttered by a man who, in his heart, thinks you a Christian dog.

though he had seen the world and twice performed the pilgrimage to Mecca, yet was a little uneasy in South Morocco, and thought it best that we should go to some caravanserai (called in Morocco fondak) and try our best to escape detection, I shamming ill, and Lutaif giving out he was a Syrian doctor.

Ali the muleteer, who learned for the first time our destination, was in an agony of fear, and said he must return at once; but when we pointed out to him that he would then not only lose his wages but perhaps his mule, he made his mind up, on condition I procured him a letter of protection from the English consul in Mogador. Mohammed el Hosein, before he left the town, had made me sign a paper stating I had engaged him for all his life, and, fortified with this protection, I understand he now bids all his governors and masters absolute defiance, wrapped, so to speak, within a tatter of the British flag. Lutaif, who knew the Governor of Tarudant, one Basha Hamou, who had been Governor of Mequinez—a negro, and a member of the famous Boukhari* Pretorian bodyguard—gave his opinion that Mohammed el Hosein was right, and that though Basha Hamou might not be pleased, he would be obliged to give us protection, and that he, probably after the first excitement of the natives had subsided, would

---

* The Boukharis were first raised by Sultan Muley Ismael, one of the most powerful rulers the country has ever had; he flourished in the eighteenth century, and sent an ambassador to.Louis XIV to demand the hand of his daughter, which, perhaps through religious intolerance, or some other reason, was not accorded to him. The Boukharis were all negroes from the Soudan, who, belonging to no Arab tribe, were devoted to the person of the Sultan alone.

send us back under escort to Mogador. I held my tongue, resolving that if we got there I would not return without a good look at the place.

About a mile from where we sat was situated the castle of the local Kaid, a castle set upon a rock, and strong enough apparently to set tribal artillery at defiance; but our Lord the Sultan being "unfavourable" to him, the castle was deserted, cattle stolen, crops all destroyed, and an air about the place reminding one of some of the Jesuit Missions (destroyed to show the Liberalism of Charles III) which I have seen in Paraguay. In fact the Kaid is only Kaid *in partibus*, and it is understood a Sheikh in Fez has offered the Sultan 100,000 dollars to be made governor, providing he (the offerer) might have a " free hand " with the tribe; this means to oppress them, and in a year or two to take the 100,000 dollars out of them to pay the Sultan, and as much more for himself. Strange that the Arabs, though so warlike, should in all ages have endured so much oppression. It may be that the tribal system renders them specially liable to this, for inter-tribal jealousies make them an easy prey to any Sultan who can command money enough to set them at one another's throats.

Ibn Jaldun (in the introduction to his history) says: " The Arabs are the least fitted men to rule other nations, for they demolish the civilization of every land they conquer. They might be good rulers, but they must first change their nature." This no doubt arises from their incapacity to govern themselves. Still with all their faults they are a fine race, and if they have demolished the civilization of several countries, they have in return left their own type wherever they have conquered, and what type in the world is finer. I say nothing of the more doubtful of their legacies, their system of numeration, and the thoroughbred horse.

The grateful spring, fruit trees, and shady little oasis where we rested rejoice in the name of Aguaydirt el Má, a compound of Arabic and Shillah, Má in Arabic meaning the water, and Aguaydirt no doubt having some meaning of its own in the wild tongue it comes from.

As we ride through a bushy country with some straggling farms, we pass a Sheikh on a good horse, long gun across the saddle, and a tail of ragged followers on foot. It seems he is a tax collector, gathering the taxes in person, and no doubt quite as effectually as the Receivers General used to do in France, including even him who sleeps (his bubble burst) under a flagstone in the door of San Moisé,* in Venice, with a brief dog-Latin epitaph, setting forth the usual lack of virtues of a man who fails.

Curious to observe (again per usual) the fatness and good clothes of the collector, and the mean estate of the poor "taxables," their downcast looks, and all the apparent shifts and wiles they are putting forth to escape the worst outrage that a free man can bear, that is to have his money taken from him under the pretext of the public good.

Pleasant to gather taxes well armed on a good horse;

* John Law of Lauriston.

a horse I mean that could do his ninety miles a day for several days and carry something heavy in the saddle bags.

I had a friend who, being for a short time governor of a province in a Central American Republic, and finding things became too hot for him, collected all the public money he could find, and silently one night abdicated in a canoe down to the coast, and taking ship came to Lutetia; and then, his money spent, lectured upon the fauna and the flora of the country he had robbed; and, touching on the people, always used to say that it was very sad their moral tone was low; the reflection seeming to reinstate himself in his own eyes, for he forgot apparently that in his abdication he but followed out the course which law had pointed out to him in his official days.

Night catches us close to el Mouerid, a dullish pile of sun-dried bricks, the lord of which, one Si Bel Arid, is an *esprit fort,* and knowing me for a Christian, ostentatiously walks up and down talking on things and others to show his strength of mind. Though disapproving of them, most Moors like to be seen with Christians, in the same way some pious men are fond of listening to wicked women's talk, not that their conversation interests, but to show the asbestos quality of their own purity, and to set forth that, as in Rahab's case no imputation can attach itself to men of virtuous life. Therefore, as maiden ladies are said to love the conversation of rakes, and clergymen, that of fallen women, so do Moors love to talk with and be seen in public with the enemies of God.

El Mouerid looks miserable in the storm of rain and wind
in which we leave it as the day is breaking. The Arab dress
in windy weather teaches one what women undergo in petti-
coats upon a boisterous day; but still their pains are mitigated
by the fact that generally men are near at hand to look at them,
whereas we could not expect to find admiring ladies on the
bleak limestone plain. Curious striations on the hills, as if
the limestone "came to grass" in stripes, give an effect as
of a building, to the rising foot-hills, into which we enter
by the gorge of Bosargun, a rocky defile which gradually
becomes a staircase like the road from Ronda to Gaucin, or
that to heaven, almost untrodden of late years. We pass
a clump of almond trees, by which a light chestnut* mare
is feeding; she looks quite Japanese amongst the trees, buried
up to the belly in aromatic shrubs; a little bird sits on her
shoulder, no one is near her, though, no doubt, some sharp-
eyed boy is hiding somewhere watching her, for in this
district no animal is safe alone. From the top we get our
first view of the Atlas in its entirety; snow, and more
snow marking the highest peaks, the Glawi, Gurgourah,
and the tall peaks behind Amsmiz. No mountain range
I ever saw looks so steep and wall-like as the Atlas; but
this wall-like configuration, though most effective for the
whole range, yet robs the individual peaks of dignity.

To the east the stony plain of Morocco, cut into channels
here and there by the diverted water of the Wad el N'fiss

---

* Light chestnut in Spanish is " ruano "; the proverb says " Caballo ruano para
las putas." Query: Does that hold good of a mare?

(a river I was destined to follow to its very source), under the highest peaks of Ouichidan, upon the very confines of the Sus. Here, for the first time, we see, though far below us, the curious subterraneous aqueducts, looking like lines of tan pits, with which the plain of Morocco is intersected everywhere.

These aqueducts, called Mitfias, are a succession of deep pits, dug at varying distances from one another; the water runs from pit to pit in a mud channel, and the whole chain of pits often extends for miles. Men who undertake such herculean labour in order to irrigate their fields cannot well be called lazy, after the fashion of most travellers who speak, after a fortnight's residence in Morocco, of the " lazy Moors." The truth is that the country-people of Morocco are industrious enough, as almost every people who live by agriculture are bound to be. Only the Arabs of the desert, and the Gauchos of the southern plains, and people who live a pastoral life, can be called lazy, though they, too, at certain seasons of the year, work hard enough. The Arabs and Berbers of Morocco work hard, and would work harder had they not got the ever-present fear of their bad government before them. When one man quarrels with another, after exhausting all the usual curses on his opponent's mother, sister, wife, and female generation generally, he usually concludes by saying: " May God, in his great mercy, send the Sultan to you "—for he knows that even Providence is not so merciless as our Liege Lord.

About three miles below us are two curious flat-topped

hills, looking like castles. Mohammed el Hosein pronounces them to have been the site of two strong castles of the Christians. What Christians, then?—Roman, or Vandal, or Portuguese? Perhaps not Christian at all, but Carthaginian; for in Morocco, any old building, the builders of which are now forgotten, is set down to the all-constructing Christians, in the same manner as in Spain, the Moors built all the castles and the Roman bridges, and generally made everything which is a little older than the grandfather of the man with whom you speak. Not but at times the person questioned puts in practice, to your cost, the pawky Spanish saying: "Let him who asks be fed with lies." What Christians could have been so foolish as to build two castles in a barren plain, far off from water, does not appear. At any rate, after a careful search, we can discover no trace of building, and put the castles down with the enchanted cities, Fata Morgana, Flying Dutchman, and the like phenomena, which seem more real than the material cities, ships, and optical illusions, which, by their very realness, appear to lose their authenticity, and to become like life, a dream.

Passing the castles, we emerged again upon a desert tract, which took almost two hours to pass, and at the furthest edge of it a zowia of a saint, Sidi Abd-el Mummen, with a mosque tower, flanked by palms, rising out of a sea of olives twisted and gnarled with age, and growing so thickly overhead that underneath them is like entering a southern church, out of the fierce glare of the sun.

History has not preserved the pious actions which caused

Si Abd-el Mummen to be canonized. In fact, Mohammed, if he came to life again, would have a fine iconoclastic career throughout the world of the Believers; for though they have not quite erected idols, graven or otherwise, yet all their countries are stuck as full of saints' tombs, zowias of descendants of saints, and adoration of the pious dead prevails as much as in the Greek or Latin churches. True, the custom has its uses, as it serves to indicate the distance on roads, and men as naturally enquire their way from Saint (Sidi) to Saint, as from church to church in Spain, or public-house to public-house in rural England. In other countries Saints, before becoming free of the fellowship, have to show their fitness for the post; but in Morocco no probation of any kind— that is, according to our ideas—seems to be necessary.

I knew an aged man, who used to sit before the Franciscan Convent, in the chief street of Tangier—a veritable saint, if saint exists. He sat there, dressed in a tall red fez,* given by some pious soldier, a long green caftan, clean white drawers, and a djellaba of fine blue cloth. Long hair descended on both sides of his face in locks like bunches of crysanthemums; his eyes were piercing, and yet wavering; for the poor Saint was nearer to Allah than the common herd by the

---

* A tall peaked fez in Morocco is the outward visible sign of a soldier or man of the Mahksen, Government. From the Arabic word Mahksen, which is not used in other Arab-speaking countries in the sense of the Government, but simply as signifying a " Store," comes the Spanish word " Almacen," a store, and some say also our word, " magazine." The inward spiritual grace is a swaggering demeanour to show the soundness of his faith, an insolence of manner not quite unknown among soldiers of other powers; and a firm determination to obtain for nothing, everything that the wretched "Pekin " has to pay for in the debased copper currency of the realm.

want of some small tissue, fibre, or supply of blood to the vessels of the brain. Thus clothed and mentally accounted for his trade, a basket by his side, and in his hand a long pole shod with iron, for he belonged to the sect called the Derkowi, he sat and told his beads, and took his alms, with an air of doing you a favour: for who gives to the poor does them no favour, but, on the contrary, insures his own eternal happiness, and but gives out again that which Allah entrusted to him for the behoof of man.

I happened one day, with European curiosity, to enquire what made the venerable man so venerated, and was told that, having suddenly gone mad, he killed his wife, threw off his clothes, and then marched naked through the land— justice not interfering—for the mad are wise; and then, the violence of his madness over, had quietly sat down and made himself a sort of "octroi" upon passers by, after the fashion of blind Bartimæus, who sat begging at the gate. The explanation pleased me, and in future when I passed I laid up treasure in that mothless territory, where no thieves annoy, by giving copper coins; and was rewarded even here on earth, for once I heard an Arab say: "This Kaffir, here" (speaking of me) "fears neither God nor devil, yet I have seen him give to the old Saint; it may be, God, the merciful, may save him yet, if but to show His might."

And so it is that Saints' tombs stud the land with oven-shaped buildings with a horse-shoe arch, a palm tree growing by, either a date or a chamcerops humilis, in which latter case pieces of rag are hung to every leaf-stalk, perhaps as an advertisement of the tree's sanctity or from some other cause. The place serves as a re-union for pious folk, for women who pray for children, for gossipers, and generally holds a midway place betwixt a church and club. In order that the faithful wayfarers, even though idiots, shall not err, in mountain passes, as in the gorge of Bosargun, at four cross-roads, passes of rivers, and sometimes in the midst of desert tracts the traveller finds a number of small cairns, in shape like bottles, which show—according to the way they point—where the next Saint's tomb lies; for it is good that man should pray and think about himself, especially upon a journey—prayer acts upon the purse; alms save the soul; and Saints, though dead, need money to perpetuate their fame.

After nine hours of alternating wind and heat we reached Imintanout, the eastern entrance of the pass which, crossing a valley of the Atlas, leads to Sus. So to speak Tarudant is within hail, three (some say two) days and we are there, if . . . but the if was destined to be mortal, as it proved. The straggling village almost fills the gorge through which the road enters the hill. Above it towers the Atlas; a little stream (then dry) ran through the place, which had an air between a village in Savoy, and a Mexican mining town lost in the Sierra Madre. Brown houses built of mud, stretches of Tabieh* walls, the tops of which crowned with dead prickly bushes, steely and bluish looking in the setting

* Tabieh, the "tapia" of the Spaniards and the "pisé" of the French, is merely mud run into frames till it hardens, and then left to dry in the sun. It figures in the saying "Sordo como una tapia," deaf as a wall, and seems to be at variance with the northern proverb, "Walls have ears."

sun, the houses generally castellated, the gardens hedged in with aloes, wherein grow blackberries, palms and pome-granates, flowers, fig-trees, and olives. Water in little channels of cement, ran through the gardens, making of them an Arab paradise. Further up the gorge the Mellah (Jewry), in which we catch a glimpse for the first time of the Atlas Jews, servile and industrious, wonderfully European-looking as to type, superior to the Arabs and Berbers in business capacity, and thus at once their masters and their slaves.

The Kaid's house, perched upon a rock, I avoided like a plague spot, fearing to be recognized and sent back to Mogador, and made, instead, for the house of one Haj Addee, a Sheikh,* which being interpreted may stand in his case for country gentleman.

The Sheikh has been in Mecca, Masar el Kahira (Cairo), carries a rosary, has some knowledge of the world (Moham-medan), and is not quite unlike those old world Hidalgos of La Mancha, they of the " Rocin flaco y galgo corredor," whom Cervantes has immortalized in the person of Don Quixote.

Friends interested in my journey in Mogador had recom-mended the Sheikh to me as a safe man in whom to trust, before engaging myself in the recesses of the Atlas. So,

---

* Sheikh is a most indefinite word, and is generally held to mean a chief, but often only means gentleman. Scribes, especially if Easterns, *i.e.*, from Syria, Damascus, or Bagdad, often use the title. Lutaif upon our journey figured as Sheikh Abdul Latif el Shami (the Syrian), and, when convenient, I was styled Sheikh Mohammed el Fasi, at other times simply " el Tabib " (the Doctor), sometimes " Sherif," anything, is fact, to distract attention from my white face and extremely small knowledge of Arabic.

riding to his house, I sent in Swani with a letter from an influential man in Mogador, and Haj Addee soon appeared, and, after asking me to dismount, led me by the hand to his guest house. This was an apartment composed of three small rooms, one serving as a bedroom, the second as a place in which to store our saddles, tent, and camping requisites, and the third, which had no roof, as sitting-room. All round the sitting-room ran a clay divan, a fire burned in one corner, and overhead the stars shone down upon us, especially the three last stars in the Great Bear's tail, so that, take it for all in all, it was as pleasantly illuminated a drawing-room as any I have seen. Hard by the door stood an immense clay structure, shaped like a water barrel, which served for storing corn* in during the winter, and in the spring broken to pieces when the corn was used.

Seated on the divan, I watched an enormous copper kettle try to boil upon a brass tripod† in which a little charcoal glowed, whilst in a small brass dish a wick fed with raw mutton fat made darkness manifest. As I look round the room it strikes me that there seems to be a sort of dominant type of Mohammedan formed by religion, in the same way that in the north of Ireland you can distinguish a Catholic from a Protestant, across the street. Mohammed el Hosein, though of a different race, and

---

* The usual system of storing grain is either in earthen jars buried in the ground, or in funnel-shaped pits known as " Metmoras," from which word the Spanish word " Mazmora " has been taken, and from which we again took our old-fashioned word " Massymore," used for a dungeon.

† This tripod is used all over Spain, and called, in Andalusia, " Anafe," from the Moorish words " En Nar fi," the fire is in it.

from thousands of miles away, presents the perfect type of an Afridi, as depicted in the columns of the illustrated papers. Ali, our muleteer, with his thin legs, beard brushed into a fan, and coppery skin, might sit for the picture of a Pathan; it may be that an Oriental would discern a great resemblance between a Dutchman and a Portuguese which had lain dormant to our faculties, and if this was the case my theory would be as well confirmed as many other theories which have revolutionized the scientific world.

We talk of Mecca and Medina, of travelling from Jeddah, stretched in " Shegedefs "* upon a camel's back, of Gibel Arafat, the Caaba, and of the multitude of different classes of Mohammedans who swarm like bees. Hindoos and Bosnians, Georgians, Circassians, the dwellers in the Straits, and the Chinese believers, whom my host serves up all in a lump as Jawi, and says that they are little, yellow, all have one face, and that their mother in the beginning was a Djin. It appears that at the sacred places, the town of tents is of such vast dimensions that it is possible to lose yourself and wander for miles if you forget to take the bearings of your tent. It must be a curious sight to see the various nationalities, the greater part of whom have no means of communication other than a few pious sentences, and a verse or two from the Koran.

Swani, who is a double pilgrim, having twice been in

---

* A Shegedef is a kind of long pannier in which the richer pilgrims lie or sit one on each side of a camel. There is an awning over all, and pilgrims have assured me that the pleasantest part of the whole journey is the portion which is performed in this manner.

Mecca, comes out most learnedly as to nice points in Mohammedan theology. Though he can neither read nor write, and is, I fear, not all too strict in the mere practice of his religion, yet he can talk for hours upon the attributes of God, and as judiciously as if he had been a graduate of St. Bees, so well he knows the essence, qualities, power, majesty, might, glory, and every proper adjective to be applied. The object of his hopes is to induce me to perform the pilgrimage. He assures me that it is quite feasible, has even arranged for my disguise, and tells me that in Mecca he can take me to a friend's house, who is as big a Kaffir as myself. His idea is that I shall go as a Circassian, which people I resemble as to type, and when I say, "What if I fall upon a real Circassian?" he only answers, "That is impossible," in the same manner as when asked what they would do if they discovered me; he answers, "they would not discover you, you look so like a man from Fez." What annoys him is that I make no apparent progress in the language, and I fear that I shall have to take a longer pilgrimage before I am fit, even with such a guide, to throw the stones on Gibel Arafat. Sometimes our talk ran on the wonders of the West; the steamships in which the pilgrims sail from Tangier to Jeddah, and on board of one of which, our host informs us, once when he was praying, the Kaffir Captain touched him on the arm, and, pointing to the compass, informed him he was not head on to the proper point. This conduct seems to have impressed Haj Addee, and he remarks, "God, in his mercy, may yet release that 'captain' from the fire." As we were talking, neighbours dropped in, in the familiar Eastern way, and sat quiet and self-contained, occasionally drinking from one of the two long-necked and porous water-jars, known as " Baradas," or the " coolers," which stand, their wooden stoppers tied to them with a palmetto cord, on each side the divan. Swani concocts the tea, using the aforesaid weighty copper kettle, a pewter cone-shaped tea-pot, made in Germany, a tin tea-caddy, painted the colour of orange marmalade, with crude blue flowers, which kind of merchandise Birmingham sends to Morocco, to be sold at one-and-sixpence, to show how much superior are our wares to those of all the world. The host knocks off great pieces from a loaf of cheap* French sugar with the key of the house, drawing it from his belt, and hammering lustily, as the key weighs about four ounces, and is eight or nine inches long.

Imintanout being, as it were, the gate of Sus, and the end of the first stage of our journey, we ask most anxiously as to the condition of the road. The way we learn is easy, so easy that trains of laden camels pass every day, and the whole distance across the mountains is a short two days. So far so good, but when we intimate our intention of starting early next morning, then bad news comes out.

It appears the tribe called Beni Sira, sons of burnt fathers, as our host refers to them, have stopped the pass, not that they are bad men, at least our host is sure of this, or lives

* It is actually sold cheaper in Morocco than in Marseilles.

too near them to venture on a criticism, but because they are dissatisfied with the new governor recently appointed, and wish to get him into bad odour with the Sultan by causing trouble. It appears those mis-begotten folk have fired upon a party only the day before, and wounded a Jewish merchant, who is laid up in a house not far from where we sit. A caravan of twenty mules was set upon last week, two men were wounded and the goods all carried off. A most ingenious system of proving that the governor is incompetent to preserve order, and therefore must be changed.

Suspecting that the story was untrue, and only got up to prevent my entering the Sus, I sent two messengers, one to see the Jewish merchant, and tell me if he is really wounded, and another to a Sheikh, asking if a traveller, going to pray in Tarudant, and skilled in medicine, can pass that way. The report of fighting seriously alarms our muleteers, and even Swani, though brave enough, looks grave at having to fight so far away from home. Haj Addee—to show goodwill, or to impress us with his power—offers, should the local Sheikh of the Beni Sira return an unfavourable reply to get his men together and fight his way right through the pass. I thank him with effusion, but resolve not to place myself alone in the middle of a tribal battle without a rifle, on a half-tired horse, and deprived even of a Kodak with which to affright the nimble adversary.

And so I lose a day, or perhaps gain it, talking to the curious people, and prescribing wisely for ophthalmia;

dividing Seidlitz powders into small portions to be taken at stated times to serve as aphrodisiacs, and watching an incantation which seems to cure our host of rheumatism.

Haj Addee was a sort of " Infeliz," as the Spaniards call a man of his peculiar temperament. I am certain that in whatever business he entered into he must have failed, he had so honest a disposition, and his lies were so unwisely gone about, they would not have deceived a Christian child. His rosary, each bead as big as a large pea, was ever in his hand; he said it was made from the horn of that rare beast, the unicorn, and he had bought it with a price at Mecca, from a " Sherif so holy that he could not lie." It looked to me like rhinoceros horn, and so, perhaps, the Sherif had lied less than he had intended when he sold the beads. In the middle of the string were four blue beads, and between these four beads a piece of ivory standing up like a cone, and called " el Madhna," * that is the " erect one." Besides the " Madhna " there was a little comb, shaped like a stable mane comb, and made of horn, which dangled from the rosary, and which Moors use to comb their beards after performing their ablutions, and which they seldom carry in Morocco, except they have performed the pilgrimage.

Haj Addee suffered from rheumatism in the left shoulder. This he called simply " el burd," " the cold," and complained that he had exhausted all lawful medicine and was

---

* " The Madhna " may be a relic of a phallic worship, many relics of which have lingered even at Mecca, as Burton, in his chapter on Mecca in his celebrated " Pilgrimage to Medinah and Mecca," relates.

about to try an incantation.* So to us entered a "fakir "—
that is, a holy man—fat and white-bearded, and with the
half-foolish, half-cunning look, so often to be found in
" holy men " who are professors of some faith either inferior
to, or differing from, our own. The man of God gave me a
scowl, and, I fancy, saw I was a misbeliever, then sat down,
after the fashion of his class, in the best seat, and, mumbling
something, drew out a dagger and wrote upon it, with the
juice of a plant he carried in his hand, some mystic characters. The master of the house, together with a friend,
stood up and held two pieces of split cane about a yard in
length. Both had the air the Italians call " compunto," that is,
they looked like people walking down the aisle of a crowded
church (as if they trod on eggs) after partaking of the Holy
Communion, and conscious that it is not impossible they
may have eaten and drunk their own damnation, and that
before a crowd of witnesses.

The Levite, sitting in a corner, kept on muttering, and
underneath the cane rapidly passed the cabalistic dagger to
and fro, just in the middle of the reeds, touching them
lightly, as the dagger moved almost like a shuttle in his
hands. The look of concentration and air of being accom-
plices after the fact was kept up for almost five minutes,
and quite insensibly I find I joined in it, and, looking at
Lutaif (a Christian, if such a man there be) at Swani and
Mohammed el Hosein, I found that they too were fascinated,

---

* I offered him quinine, but he looked coldly at it as a man in the time of
Molière might have disdained the futile drugs of the licensed practitioner knowing
that he had orvietan at his command.

much like a rabbit in a snake-house moves towards the snake.

It seems, of all the forces which move mankind, humbug is the strongest, for humbugs are always taken in by humbug, and the very men who practise on the folly of mankind fall easy victims to the manœuvres of their brothers in the art. Gradually the movement of the dagger grew slower, then stopped, and then the mystery man struck the patient gently with the blade upon the arm, broke the two canes across, tied them together in a bundle, and put them on a ledge just where the thatch is fastened to the mud walls of the house.

Admirable to observe the look of faith of all the standers-by. If ever I witnessed a religious action, that is, taking religio and superstitio (as I think the Romans did) to be synonymous, I did so then. Even the Salvation Army, at its most unreasonable, never succeeded in bringing such a look of ovine faith upon the faces of its legionaries. The patient says that he is greatly benefited, and for the first time in my life I see a religious ceremony begun, continued, and concluded (with the result successful), and without apparent sign of a collection being taken up.

Faith should not be divorced from its true spouse, the offertory. What man has brought together, even the act of God should not disjoin, for faith is so ethereal that, left alone, it pines and languishes in a hard world, unless sustained by pennies in a plate, in the same manner as the soul, which, as we all know (nowadays), is eternal, takes its departure when the body dies, and lies unquickened if the head receives a knock.

The function over, we sit and talk, look at one another's arms, lie as to our shooting powers, and, after careful examination of my pistol, to my astonishment, our host produces a large-sized Smith and Wesson revolver, which it appears he bought in Kahira (Cairo), and seems to think the most important result of all his pilgrimage. Hours pass, and still no messenger, and a fat Saint* rolls in, and salutes Swani as an acquaintance of the pilgrimage. They kiss and embrace, hugging one another and alternately placing their beards over one another's right and left shoulders. As Burton observes, the Arabs are by far the most emotional of the Oriental races. Poetry, or a well-told lamentable tale, moves them to tears, and friends who meet often weep for joy.

Swani privately imparts to me, in pidgin-Spanish, which he tells the Saint is Turkish, that the Saint is a humbug, and that when they last met he would hardly speak to him, for he (Swani) was working as a sailor on board an English pilgrim steamer, and dressed "a la inglesa, saber del trousers, catchy sea boots y todo."

Still waiting for the messenger, I fall to observing the difference between the Arabs and the Berber race. It is the fashion amongst travellers in Morocco and Algeria to exalt the Berbers, and run down the Arabs. "The noble Shillah race" has become quite a catch-word with every one who

---

* The word Sherif is often rendered by Europeans, in Morocco, as Saint, they having most likely taken the word from the Spanish word Santon.

sees a Berber and writes down his impressions. Curious that no one talks of the noble " Tuareg race," yet the Tuaregs are Berber of the Berbers, and their language, known as Tameshek, merely a dialect of the Berber language, which spreads over the vast area of territory from Tripoli to the Atlantic Ocean, and from the Mediterranean to Timbuctoo. Undoubtedly the oldest known inhabitants of the countries which now are called the Barbary States, they seem to have kept their type and customs unchanged since first we hear of them in history. The Arabs found them in possession of all Morocco, and drove them into the mountains and the desert beyond, and though they forced Islam upon them, still the two peoples are anti-pathetic to one another, have blended little, and you can tell them from one another at a glance.

The Arab, one of the finest types of all the races of mankind, tall, thin, fine eyes, aquiline noses, spare frames; walking with dignity; a horseman, poet; treacherous and hospitable; a gentleman, and yet inquisitive; destroying, as Ibn Jaldun assures us, the civilization of every land they conquer, and yet capable of great things—witness Granada and Damascus; a metaphysician and historian; sensual and yet abstemious; a people lovable, and yet not good to " lippen to," as Scotchmen say; and yet perhaps of all the Orientals those who have most impressed themselves upon the world.

The Berbers, short, squat figures, high cheekbones, small eyes, square frames, great walkers, only becoming horse-

men by necessity, as when the Arabs have forced them to
the desert; as fond of mountains as the Arabs are of plains;
in general agriculturists, whereas the Arab in his true sphere
prefers a pastoral life; the Berbers, little known outside
their mountains, look rather Scottish in appearance, that is,
Scotch as ordinary mortals see that race, and not as seen
through "kailyard" spectacles. It may be that the Berbers
are a noble race, but personally I should apply the adjective
"noble" to the Arabs, and to the Berbers give some such
qualifying phrases as "relatively honest," "tenacious," or
perhaps, best of all, "bourgeois," which, to my way of
thinking, best expresses the characteristics of those Berber
tribes who, in the north side of the Atlas, are dominated by
the Sultan's power. Where they touch one another to the
south upon the confines of the desert, it would be hard to
give the palm for savagery; and the great Berber tribes, Ait
Morghed and Ait * Hannu, have become practically Arabs
in their customs and their use of horses. The Tuaregs,† on
the other hand, have remained absolutely Berber, and in-
discriminately attack Arabs and Christians, and all who
cross their way.

* Ait corresponds to the Arab "Ibn" or "Ben," and the Scotch "Mac."

† It was the Tuaregs who killed the French explorer, the Marquis de Morès, and
they have killed many explorers of almost every European nation. From their
habit of going veiled to protect themselves against the sun and dust in the desert,
some have supposed that the mysterious "veiled men" referred to in the Spanish
Chroniclers as having accompanied the Almohades in their invasion of Spain in
1146, were Tuaregs. The leader of the Almohades was Mohammed-ibn-Abdullah,
King of Fez, or Morocco as some say, for the kingdoms were not joined in those
days. In either case, he might have brought the Tuaregs. The word Almohade is
said to mean Unitarian, a title of honour in lands where miscreants either reject
or do not fear the doctrine of the Trinity.

As to the name of Berber, ethnologists, after the fashion of all scientists, have disagreed, taken one another by the beards, and freely interchanged (and I suppose as faithfully received) the opprobrious names which render the disputes of men of science and of theologians so amusing to those who stand aside and put their tongues out at both sorts of men.

Breber, Baraba, Berber, all three phases of the word are found. Some learned men derive the word from the Greek, and make it simply stand for Barbarian. Others, again, as stoutly make it Arab, and say it means " People of the Land of Ber."*

Ibn Jaldun (always an innovator) has his theory, which seems just as good as any other man's. He makes the Berbers to descend not from Shem, but Ham (Cam as the Arabs call him), and relates that Ham had a son called Ber, whose son was called Mazirg, and that from Ber came the Beranis, who, in time, and by corruption of the word, turned into Berberes. If not convincing, the theory shows invention, and smacks (to me) of the derivations I have heard hacked out, so to speak, with a scalping knife round the camp-fire; for, uncivilized and semi-civilized men waste as much time in seeking to find out how the form of words got crystallized as if they had diplomas from their universities.

Strangely enough, the people we call Berbers do not know the name, and call themselves Tamazirght, that is, the noble. Their language is called Amzirght, and resembles

* Gråberg di Hemsö. " Specchio Geographico e Statistico di Marocco," page 72.

closely the Tamashek spoken by the Tuaregs, the dialect of
the tribes of the Riff Mountains, and that of the Kabyles of
Algeria.* The Arabs neither use the word Tamazirght nor
the word Berber, but call the Berber tribes " Shluoch," that
is, the outcasts; the verb is " Shallaha," and the term used
for the speech Shillah,† a sort of Shibboleth in Europeans'
mouths, for very few even of those who, like the Germans
and the Spaniards, can pronounce the guttural, ever attain
to the pronunciation of the Arabic and Berber semi-guttural,
semi-pectoral aspiration of this word.

   Till the last twenty years the greater portion of the Berber
tribes, although Mohammedans, owned but a nominal
allegiance to the Sultan of Morocco, and lived almost inde-
pendently under their Omzarghi (lords), Amacrani (great
one), and Amrgari (elders), but nowadays, even in Sus, and
in the desert, the Sultan's authority is much more felt. In
fact, until quite lately they lived as their forefathers—the
Getulians, Melano-getulians, and Numidians—lived before
them, with the exception that being driven to the mountains

---

* For list of Amzirght words collected at Gundaffi, see notes.
   † Mr. Walter Harris, in his Tafilet, says the word Shillah = noble, but he has
probably been informed by a Berber. He also, after the fashion of most European
travellers, " finds out immediately how infinitely superior they are (the Berbers)
in morals and character to Arabs. Their every word and look speak of greater
honesty and truth than one finds in a month amongst the Arabs." ("Tafilet,"
p. 62.) Certainly Mr. Harris has every right to speak, as few men know the
mountaineers better than he does, and dressed in their clothes, his head shaved,
and a string of camel's hair bound round his forehead, bare feet and legs, and
wrapped in a brown djellaba, he could pass anywhere for one of those moral and
honest folk. I wish, though, that he had stated plainly what he understands by
Shillah honesty and morality, for, as in theological discussions, the greatest
difficulty is to define terms.

they had greatly lost the horsemanship which made them famous to the ancient world. Even today it is not rare to see them ride, just as the Roman writers said the ancient Numidian cavalry rode, without a saddle or a bridle, and guiding their horses with a short stick, which they alternately change from hand to hand to make them turn.

The words Mazyes, Maziriciæ and Mazyces occur in many Greek and Roman writers, and seem not impossibly to be derived from Amazirght, or from Mazig, the most ancient form known of their appellation. Leo Africanus, himself a Moor,* calls them Amarigh, and says of them " they are strong, terrible, and robust men, who fear neither cold nor snow; their dress is a tunic of wool over the bare flesh, and above the tunic they wear a mantle. Round their legs they have twisted thongs, and this serves them also for shoes. They never wear anything on the head at any season; they rear sheep, mules and asses, and their mountains have few woods. They are the greatest thieves, and traitors, and assassins in the world——"†

Even today this picture of them holds good in most particulars. The Berbers of the mountains seldom wear turbans or anything but a string tied round their foreheads. Generally they have a linen shirt today, but often wear the tunic as Leo Africanus says, and now and then one sees them with

---

* He was born in Granada, and fled to Fez after the capture of Granada by the Catholic Kings. Being taken prisoner by Christian pirates, he was brought to Rome, received into favour by the Pope, was baptized, and died at Rome after translating his work on Africa into Italian.

† " Noble Shillah race " of modern travellers. " Moral and honest " folk of Mr. Harris, etc., etc.

the twisted leg bandages like Pifferari. As to their moral character, after some small experience, I rather hold to the view of Leo Africanus, than that of Mr. Walter Harris. One thing is certain, that they cannot lie more than the Arabs do, but then the Arabs lie so prettily, with so much circumstance, and such nice choice of words, that it all comes to be a matter of individual taste, for there are those who had rather be deceived with civil manners by an Italian, than be cheated brutally by a North Briton, for the love of God.

The first of all the tribes we hear of in history as living in Morocco are the Autoloti of Ptolemy, who seem to be the Holots, who now live in El Gharb, that is, the country between Tangier and the Sebu. In Hanno's Periplus, the same word occurs, and the description of their country seems to tally with the territory where they live.

Luis de Marmol, who was long prisoner with the Moors in the sixteenth century, places a people called Holots, near to Cape Azar. Now it is certain that the Holots are Berbers, and the testimony of the writers referred to goes to prove the long continuance of the Berbers in the land, and also seems to prove that in ancient times, as now, the Berbers were not nomads, as the Arabs were, but stationary, as they are today. Gråberg di Hemsö, in his curious " Specchio Geographico di Marocco," says, " Questi Mazighi della Tingitana fabbricarono, ne quella costa in vincenanza del capo Bianco la citta di Mazighan, che porta ad oggi il loro nome di nazione." He quotes no authority for his statement, and it is certainly at variance with fact, for Mazagan was founded by the Portuguese in 1506, and it is called Djedida, *i.e.*, New Town, by the Arabs; still the name of Mazagan may yet commemorate an older town under the style of Mazighan.

Be all that as it may, the Berbers inhabited Morocco ages before the Arabs conquered the land, and gave them the religion of the sword. Tradition says they were once Christians, and certainly in their embroideries and decorations the cross is used. Yet, looking at the matter from an artistic point of view, discounting (for the nonce) morality, the cross, and honesty, it seems to me that noble is not a term to use in speaking of the Berber, and I submit it better fits a race such as the Arabs, who, in the persons of their horses and themselves, have done so much to refine the type of all those peoples, equine and biped, whom they have come across.

As night was falling, and a whistling wind springing up from the mountains, the messenger appeared bringing the news that the Sheikh refused to let anyone cross the pass, and that there had been a pretty hot exchange of shots that afternoon between the men of the Beni Sira, and a troop of cavalry crossing from Sus. This was a stopper over all, for it was evident we could not force our way through a road winding by precipices, held by a mountain tribe. Remembering the Arab proverb which runs, " Always ask counsel from your wife, but never act on what she says," I held a long palaver with Haj Addee, Lutaif and my men, on what was best to do. Two courses were now open to me, either to

wheel about and follow the shore road by Agadir, or else
to try the upper and more difficult pass to Sus, which starts
above Amsmiz. This place was situated about two days'
journey further on, the pass, according to all reports, took
three good days to cross, and having crossed it, we should
still be a long day from Tarudant.

Lutaif had no opinion, and as the rest all counselled
Agadir, except Haj Swani, who gave it his opinion that he
would go wherever I did, I " opted " for Amsmiz, arguing
that to turn back would certainly dishearten my companions, and if the Howara had been fighting a week ago,
they would be fighting still, and thinking that even if taking
the upper road I failed, I should see more of the interior of
the Atlas, than I was likely to do by any other route. After
having cursed the Beni Sira thoroughly in all the languages
we knew, drunk gallons of green tea, sat for an hour or two
listening to stories of the Djinoun, smoked cigarettes and
Kiff, and generally tried to imagine we were not disappointed, we retired to bed, so as before first light to be
upon the road. Our bedroom had no window, and gave on
the *al fresco* drawing room I have referred to; all round the
walls were little recesses in which to put things, made in the
thickness of the wall, pouches and powder horns hung
from goats' horns forced underneath the thatch, three long
" jezails," all hooped with silver—one with a Spanish two-
real piece depending from the trigger-guard—stood in the
corner, a lantern made of tin with coloured glass gave a

red light, upon the floor of mud a Rabat carpet in pattern like a kaleidoscope or Joseph's coat was spread; nothing of European manufacture was there except a large-sized (navy pattern) Smith and Wesson pistol, which, hanging by a red worsted cord upon the wall, seemed to project the shadow of the cross upon the room.

Chapter IV

MY JOURNEY all next day lay through low hills of reddish argillaceous earth, cut into gullies here and there by the winter rains, and clothed with sandaracs, suddra (Zizypphus Lotus) and a few mimosas. The hills sloped upwards to the wall-like Atlas, and on the left the desert-looking plain of Morocco, broken but by the flat-topped hill known as the Camel's Neck (Hank el Gimel), and bounded by the mountains above Demnat, and the curve the Atlas range makes to the north-east so as to almost circle round Morocco city. We search our hearts, that is, I search my own and the others' hearts and try to persuade myself that we have acted prudently in not attempting the Imintanout road. But when did prudence console anybody? Rashness at times may do so, but the prudent generally (I think) are more or less ashamed of the virtue they profess. The Moors, of course, were glad we had risked nothing, for though no cowards when the danger actually presents itself, indeed, in many cases (as at the battles with the Spaniards in 1861) showing more desperate courage than anyone except the Soudanese, still they are so imaginative, or what you please, that if men talk of danger they will take a five days' journey to avoid the places where hypothetic peril lurks. Mohammed el Hosein, being in his character of ex-slave dealer, what the French call a " lapin," talked of his adventures in the past, told how he had smuggled slaves into the coast towns almost before the Christian consul's eyes, sung Shillah songs in a high, quavering falsetto, and boasted of his prowess in the saddle, occasionally bursting into a suppressed chuckle at the idea of taking a Christian into Tarudant.

Almost before we were aware of it, on going down a slope between some bushes, we found ourselves right in the middle of a crowded market. These country markets are a feature of Morocco, and, I think, of almost every Arab country. Often they are held miles away from any house, but generally on an upland open space with water near. When not in use they reminded me of the " rodeos," on to which the Gauchos, in La Plata, used to drive their cattle to count, to mark, part out, or to perform any of the various duties on an " estanciero's " * life. The markets usually are known by the day of the week on which they are held, as Sok el Arba, Sok el Thelatta, el Jamiz, and so forth; the Arabs using Arba, Tnain, Thelatta, " one, two, three " etc., to designate their days. This market, in particular, I knew to be the Sok es Sebt, but thought we had been some distance from it, and all my assurance was required to make my way amongst at least two thousand people with the

* Estancia is the Argentine term used to denote a cattle farm. In Spain it is rarely or never used in such a sense, and the word Cortijo is the usual term for a farm.

dignity which befits a Moorish gentleman upon a journey. Arabs and Berbers, Jews and Haratin (men of the Draa province, of mixed race) were there, all talking at the fullest pitch of strident voices, all armed to the teeth. Ovens with carcases of sheep roasting entire inside them, cows, camels, lines of small brown tents made to be packed upon a mule and called "Kituns," dust, dust, and more dust, produced the smell, as of wild beasts, which emanates from Eastern crowds. Moors from Morocco city in white fleecy haiks I carefully avoided, as being my equals in supposititious rank, and, therefore, likely to address me. Berbers in striped brown rags, and wrapped in the curious mantle, called an "achnif" in Shillah, made of black wool with fringes and an orange-coloured eye, about two feet in length, woven into the back, abounded; through them I shoved my horse, not even looking down when the poor fellows lifted my cloak and kissed the hem, and, passing through majestically, I heard some mutter, "That Sherif is very proud for one so thin," fat being amongst Moors a sign of wealth, as it was evidently amongst the Jews, if we are to take the testimony of the Old Testament as worthy of belief.

To ride right through a market and pass on would have looked suspicious, as markets in Morocco form a sort of medium for exchange of news, in the same manner that in old times the churchyard was a kind of club; witness the story of the elder who was heard to say—"he would not give all the sermons in the world for five good minutes of the churchyard clash." So, after having gone about a quarter of a mile, we got under some olive trees (zeitun, from whence the Spanish aceituna, an olive), and, sitting in the shade, sent Swani to the market to buy some " Schwah," that is, some of the carcase of a sheep roasted whole " en barbecue." Various poor brothers in Mohammed came to assure themselves of my complete good health; but Mohammed el Hosein informed them the Sherif was ill, and, giving them some copper coins, they testified to the existence of the one God, and hoped that He might in His mercy soon make me well. Being schooled as to the form to be observed, I looked up slowly, and, raising one hand, muttered as indistinctly as I could that God was great, and that we all were in His hands. This pleased Mohammed el Hosein so much that when we were alone he assured me I must have been born a Sherif, and could I but speak Arabic a little less barbarously that our journey would have been productive, as Sherifs make it a practice, whilst giving small copper to the poor, to cadge upon their own account from the better classes, of course for Allah and His Holy Prophet's sake. Lutaif, who in his character of Syrian, talked almost incessantly to anyone we met, elicited that this particular market called " Sok es Sebt," the " Sixth Day Market," is one of four held in Morocco on a Saturday, thus showing that the Jews have got most of the trade into their hands, and do not care that markets should be held upon their holy day. We tore the " Schwah " between our fingers, in the " name of God "; it tasted much like leather cooked in suet; we passed the ablutionary water over our hands, moving them

to and fro to dry, drank our green tea, and were preparing
for a siesta when it was rumoured that an English Jew was
soon expected to arrive. Having still less acquaintance with
Yiddish than with Arabic, and being certain that the English Israelite would soon detect me, and fall on one shoulder
exclaiming " S'help me, who would have thought of meeting a fellow-countryman out 'ere !" I saddled up and started
as majestically as I felt I could upon my way. The start was
most magnificent, Swani and Mohammed helping to arrange
my clothes when I had clambered to my seat, and all went
well with the exception that no one happening to hold
Lutaif's off stirrup, and the huge Moorish mule saddle,
called a " Sirijah," being so slackly girthed, that it is almost
impossible to scale it all alone, he fell into the dust, and Ali
coming up to help with a broad grin received a hearty
" Jejerud Din! " (" Curse your religion! "), which caused
a coolness during the remainder of the journey. Lutaif,
though a good Christian and as pious as are most dwellers
on the Lebanon, yet came from the country where, as Arabs
say, " the people all curse God," referring to the imprecation on their respective creeds, which is most faithfully taken
and received between Maronite and Druse, Christian and
Moslem, and all the members of the various jarring sects
who dwell under the shadow of the cedars on those most
theologic hills. Nestling into the gorges of the hills, and
crowning eminences, were scattered villages of the true Atlas
type, built all of mud, flat roofed, the houses rising one over
the other like a succession of terraces or little castles, and

being of so exactly the same shade as the denuded hills on
which they stand as to be almost impossible to make out till
you are right upon them. At times so like is village to hill,
and hill to village, that I have taken an outstanding mound
of earth to be a house, and many times have almost passed
the village, not seeing it was there.

Riding along and dangling my feet out of the stirrups to
make the agony of the short stirrup leather, hung behind the
girths, endurable, it struck me what peaceful folks the Arabs
really were. Here was a traveller almost totally unarmed,
for the Barcelona and Marseilles " snap-haunces " we had
borrowed at the Palm-Tree House were hardly to be called
offensive weapons; without a passport, travelling in direct
defiance of the Treaty of Madrid, in Arab clothes, asserting
that he was an Arab or a Turk (as seemed convenient). So,
unattended, for defensive purposes, I rode along quite
safely, except from all the risks that wait upon the travellers
in any country of the world, such as arise from stumbling
horses and the like, fools, and the act of God.

What was it that stopped a band of Arabs on the look-out
for plunder ? What stood between us and a party of the
" Noble Shillah race " ? Either or both could easily have
plundered us and thrown our bodies into some silo and no
one would have known. The truth is, in Morocco, when
one reflects upon the inconveniences of the country, the
lonely roads, the places apparently designed by Providence
to make men brigands, and the fact that almost every Arab

owns a horse and is armed at least with a stout knife, that the inhabitants are either cowardly to a degree, are law-abiding to a fault, or else deprived by nature of initiative to such extent as to be quite Arcadian in the foolish way in which they set about to rob. When I remembered Mexico not twenty years ago, before Porfirio Diaz turned the brigands, who used to swarm on every road, into paid servants of the Government, men on a journey from the Rio Grande to San Luis Potosí all made their wills and started weighted down with Winchesters, pistols and bowie-knives, besides a stout machete stuck through their saddle-girths. So Morroco seemed to me a perfect paradise.

In the republic of the Eagle and the Cactus, the tramways running to the bullfights at Tacubaya frequently were held up by armed horsemen and the passengers plundered of everything they had about them. Stage coaches often were attacked by "road agents" and everyone inside of them stripped naked, although the robbers, being caballeros, generally distributed some newspapers amongst them to cover up their nakedness. In the old monarchy, murder was pretty frequent, but it was chiefly the result of private vengeance, and though the tribes fought bloody battles now and then, a travelling stranger seldom was molested by them, except he got in the way.

Real highway robbery seemed not to flourish in Morocco, perhaps because the atmosphere of monarchy was less congenial to it than the free air of a republic, though that could surely not have been the case, as virtue is a plant that grows from the top downwards, and both the President and Sultan, at the time I write about, are robbers to the core.

Occasionally in the Shereefian Empire, a Jew returning from a fair was set upon and spoiled, and now and then in lowered voices, as you jogged along the road, were pointed out to you the place where Haramin * had slain a man thirty or forty years ago.

This is the case in all those portions of Morocco where Christians travel; that is to Fez, to Tetuan, from Mazagan to the city of Morocco, and generally about Tangier, the coast towns and the Gharb.† Outside those spheres the case is different; and in the Riff, the Sus, Wad Nun, or even a few miles outside of Mequinez, a Christian's life, or even that of a Mohammedan from India, Persia or the East, would not be worth a " flus." ‡ We rested for our mid-day halt upon the open plain under the shadow of a great rock, the heat too great to eat, and passed the time smoking and drinking from our porous water-jars, until the " enemy," as the Arabs call the Sun, sank down a little; then in the cool we went on through a wilderness of cactus and oleanders,

* Literally, Bastards.

† El Gharb is a prairie territory stretching from Tangier to the river Sebu. The word Gharb means " the west," and Algarve in Portugal was simply the west of that country. The word Trafalgar is compounded from Tarf, a headland, and El Gharb, the west.

‡ Flus is a small copper coin, a donkeyload of which about makes change for a sovereign. It has come in Morocco to mean money generally, and was evidently so used in Spain under the Moorish domination, for I remember seeing a coin the inscription on which was " this flus was coined in Andalous "—i. e., Spain, which the Moor generally referred to as " Andalous."

almond and fig trees, with palms and apricots, till sandy paths zigzagging between aloe hedges with a few tapia walls backed by a ruined castle, betokened we were near Asif-el-Mal. Asif means river in Shillah, and at the foot of the crumbling walls a river ran, making things green and most refreshing to the eyes after ten hours of fighting with the enemy, enduring dust, and kicking at our animals after the fashion which all men adopt upon a journey; not that it helps the animals along, but seems to be a vent for the impatience of the traveller. Norias * creak, a camel with a donkey and a woman harnessed to one of them, water pours slowly out of the revolving " Alcuzas," † and at the trough the maidens of the village stand waiting to fill their water-jugs, shaped like an amphora, which they carry on the shoulders with a strap braced round their hands. Asif-el-Mal boasts a Mellah, and Jews at once came out to offer to trade with us, to talk, and hear the news. Intelligent young Jews, Moisés, Slimo, and Mordejai with Baruch, and all the other names, familiar to the readers of the Old Testament, flock round us. All can read and write, can keep accounts, as well as if they had been born in Hamburg; most have ophthalmia, some are good-looking, pale with great black eyes, and every one of them seems to be fashioned with an extra joint about the back. Yet hospitable, civil and a link

* Noria is the Persian water-wheel, Naurah in Arabic, which literally means a machine, and as it probably was the greatest machine at the time of its invention (say B. C. 5,000), the name has remained.

† Alcuza is an earthenware jar, fixed to the water-wheel, which empties itself as the wheel turns round.

with Europe, which they have never seen; but about which they read, and whose affairs they follow with the keenest interest, all knowing Gladstone's name and that of Salisbury; all longing for the day when they can put on European clothes and blossom out arrayed after the gorgeous fashion of the tribe in Mogador. Each of them wears a lovelock hanging upon his shoulder, and, without doubt, if they were but a little bit more manly-looking, they would be as fine young men as you could wish to see.

Their race almost controls the town, Berbers are few and hardly any Arabs but the Sheikh and his immediate following live in the place. A proof the land is good, the soil productive, and the water permanent; for when did Jews set up their tabernacles on an unproductive soil, in a poor town, or follow the fortunes of any one who was not rich? Their chief, Hassan Messoud, a venerable man, dressed in a long blue gown, a spotted belcher pocket-handkerchief over his head and hanging down behind in a most unbecoming style, advanced to greet us. Perhaps he was the finest Eastern Jew I ever saw; a very Moses in appearance, as he might have been on Sinai a little past his prime, and yet before the ingratitude of those he served had broken him. Beside him walks his daughter, a Rebecca, or Zohara, bearing fresh butter in a lordly tin dish, and bread baked upon pebbles, with the impression of the stones upon the underside. Such bread the chosen people have left in Spain, and still in Old Castille amongst the " Cristianos Rancios," who hate the very name of Jew, and think that the last vestige of their customs has long left Spain, the self-same bread is eaten at Easter, if I remember rightly; and so, perhaps, the true believer (Christian this time) unwittingly bakes bread which yet may damn him black to all eternity.

Hassan comes quickly into the tent, and bids us welcome in Jewish Arabic; and waiting cautiously till even our own Moors have left the tent, breaks into Spanish and asks me of what nationality I am. I tell him from " God's country," and he says, " Ingliz," to which I answer, " Yes, or Franciz, for both are one." He grins, squatting close to the door of the tent, servile, but dignified, full six feet high, his black beard turning grey, and hands large, fat, and whitish, and which have never done hard work. We talk, and then Hassan takes up his parable. Glory to Allah he is rich, and all the Governors are his dear friends, that is they owe him money; and as he talks the old-time Spanish rhyme comes to my memory, which, talking of some Jews who came to see a certain king, speaks of their honeyed words, and how they praised their people and boasted of their might.

> " Despues vinieron Don Salomon y Don Ezequiel,
> Con sus dulces palabras parecen la miel,
> Hacen gran puja, de los de Israel."

And yet Hassan was but easy of dispense, wearing the clothes of an ordinary Morocco Jew, with nothing to indicate his wealth.

As we sat talking to him of the exchange in Europe;

about the Rothschilds, Sassoons, Oppenheims, and others of
the " chosen " who bulk largely in the money columns of
the daily press of Europe (all of whom he knew by name),
and interchanging views about the late Lord Beaconsfield,
whom Hassan knew as Benjamin ben Israel, the daughters
of God's folk came out upon the house-tops, dressed in red
and yellow, with kerchiefs on their heads, eyes like the
largest almonds, lips like full-blown pomegranates, and
looked with pride upon their headman talking to the Sherif
or Christian Caballer, whichever he might be, on equal
terms. Strange race, so intellectual, so quick of wit, so subtle,
and yet without the slightest dignity of personal bearing;
handsome, and yet without the least attraction, conquering
the Arabs as they conquer Saxons, Latins, or all those with
whom they came in contact upon that modern theatre of
war—the Stock Exchange. Hassan Messoud having pro-
tested that by the God of Abraham we were all welcome,
retired and left us to pitch our tent upon a dust-heap in a
courtyard, between some ruined houses and the village wall.

Under the moonlight, the distant plain looked like a vast,
steely-blue sea, the deep, red roads all blotted out, the palms
and olives standing up exactly as dead stalks of corn stand
up in an October wheatfield. The omnipresent donkeys and
camels of the East hobbled or straying in the foreground
beneath the walls, and the mysterious, silent, white-robed
figures wandering about like ghosts, the town appeared to
me to look as some Morisco village must have looked in
Spain when the Mohammedans possessed the land, and

villages brown, ruinous, and hedged about with cactus like Asif-el-Mal, clung to the crags, and nestled in the valleys of the Sierra de Segura, or the Alpujarra.

At daybreak, Hassan Messoud appeared with breakfast at our tent, olives and meat in a sauce of oil and pepper, not appetising upon an empty stomach, but not to be refused without offence. Then, throwing milk upon the ground from a small gourd, he blessed us, and invoked the God of Israel to shield us on our way. A worthy, kindly, perhaps usurious, but most hospitable Israelite, not without guile (or property), a type of those Jews of the Middle Ages from whom Shakespeare took Shylock, and in whose hands the lords, knights, squires, and men-at-arms were as a Christian stockbroker, cheat he as wisely as he can, is today in the hands of any Jew who, a few months ago, retailed his wares in Houndsditch; but who, the Exchange attained to, walks its precincts as firmly as it were Kodesh, and he a priest after the order of Melchizedec. Riding along the trail which runs skirting the foothills of the Atlas, and forces us to dive occasionally into the deep dry " nullah," for there are only six or seven bridges in all Morocco, and none near the Atlas, the vegetation changes, and again we pass dwarf rhododendrons, arbutus, and kermes-oak, and enter into a zone of plants like that of southern Spain, with the exception that here the mignonette becomes a bush, and common goldenrod grows four feet high, with a thick woody stem. White poplars, walnuts, elms, and a variety of ash are planted round the houses. From the eaves hang strings of maize cobs; bee-hives like those the Moors left in Spain, merely a hollow log of wood, or roll of cork, lie in the gardens; grape vines climb upon the trees, producing grapes, long, rather hard, claret-coloured, and aromatic, the best, I think, in all the world, and which have fixed themselves upon the memory of my palate, as have the oranges of Paraguay.

About midday, upon a little eminence, we sight the tower of the Kutubieh, the glory of Morocco; but the city is, so to speak, hull down, and the white tower seems to hang suspended in the air without foundations; indeed, it looks so thin at the great distance from which we see it, as to be but a mere white line standing up in the plain and pointing heavenwards, that is, if towers built by false prophets do not point to hell. All day my horse, really the best I ever rode in all Morocco, had been uneasy, wanting to die down; so disregarding the advice of Mohammed el Hosein to prick him with my knife and pray to God, somewhat prosaically I got off, and, unsaddling him, found his shoulder fearfully swollen, and understood how I came possessed of a horse the like of which few Christians, even for money, get hold of from a Moor. I saw at once he had a fistulous sore right through the withers, incurable without an operation and a long rest. Then I believed what Mohammed el Hosein had told me, that the horse was from the other side of the Atlas and had been used in ostrich-hunting, for a better and more fiery beast I never rode, and though thin, rough, and in the worst condition, had he been sound-backed, quite fit to carry me to Timbuctoo.

The Arabs have an idea upon a journey that a man should dismount as seldom as he can. They say dismounting and remounting tires the horse more than a league of road, they therefore sit the livelong day without dismounting from their seat. The Gauchos, on the contrary, say it helps a horse to get off now and then and lift the saddle for a moment, loosening the girth so that the air may get between the saddle and the back. This the Moors hold to be anathema, and Spaniards, Mexicans, and most horsemen of the south agree that the saddle should not be shifted till the horse is cool, on pain of getting a sore back. Who shall decide when horsemen disagree? However, the Gauchos all girth very tightly, and the Arabs scarcely draw the girth at all, their saddles repose on seven (the number is canonical) thick saddle cloths, and are kept in their place more by the breast-plate than the girth. It may be therefore that, given the loose girth, short stirrup leathers, and their own flowing clothes, personal convenience has more to do with the custom of not getting off and on than regard for the welfare of their beast. The Arabs in Morocco, though fond of horses, treat them roughly and foolishly; at times they cram them with un-necessary food, at times neglect them; their feet they almost always let grow too long, their legs they spoil by too tight hobbling, and if upon a journey their horse tires they ride him till he drops.

Across the mountains and amongst the wilder desert tribes, this is not so, and travellers all agree that the wild Arab really loves his horse; but he has need of him to live, whereas inside Morocco horses are used either for war or luxury, or because the man who rides them cannot afford a mule. The pacing mule, throughout North Africa, is as much valued as he was in Europe in the Middle Ages, and commands a higher price than any horse; yet whilst allowing that he is dogged as a Nonconformist, on the road, sober and comfortable, to my eye, a man wrapped in a white burnouse, perched on a saddle almost as large as the great bed of Ware, looks fitter to be employed to guard a harem than to enjoy the company of the houris inside. Necessity (in days gone by) has often forced me to ride horses with a " flower " * on their backs or with a sore which rendered every step they made as miserable to me as it was hell to them; but as on these occasions I rode either before an Indian Malon,† or for the dinner of a whole camp, so I at once determined that, notwithstanding any risk, I would purvey me a new horse at the next stopping place. As it turned out, the changing horses and the talk which ensued, as the owner of the horse and I tried to deceive each other about our beasts, was the occasion of my never reaching Tarudant. At least I think so, and if it was, some better traveller will

* The Gauchos used to call a sore on a horse's back " una flor," a flower, and they certainly rode their horses no matter how red the " flower " was, as if their own withers were unwrung.

† Malon was the word used on the Pampa to designate an Indian invasion. I put it to casuists if it was permissible on these occasions to ride a sore-backed horse, and still be called a humane man.

do that which I failed in doing, write a much better book than any I could write, and if he be a practical and pushful man spare neither horses on the road, nor stint the public of one iota of his " facts," for to the pushful is the kingdom of the earth. As to what sort of kingdom they have made of it, it is beyond the scope of this poor diary to enquire.

We crossed the Wad el Kehra, and early in the afternoon tied up our animals under a fig-tree, with a river running hard by, a stubble field in front, and Amsmiz itself crowning a hill upon our right. Amongst the " algarobas,"* fig-trees and poplars, swallows flit, having come south, or perhaps migrated north from some more southern land.

At the entrance of the town stood the palace of the Kaid, an enormous structure made of mud and painted light rose-pink, but all in ruins, the crenellated walls a heap of rubbish, the machicolated towers blown up with gunpowder. The Kaid, it seems, oppressed the people of the town and district beyond the powers of even Arabs and Berbers to endure; so they rebelled, and to the number of twelve thousand besieged the place, took it by storm, and tore it all to pieces to search for money in the walls.

Most people in Morocco if they have money, hide it in the walls of their abode, but the Kaid of Amsmiz was wiser, and had sent all his to Mogador. He fought to the last, then cutting all his women's throats, mounted his favourite horse and almost unattended " maugre all his enemies, through

---

* " Algaroba " is one of the words the Spaniards have taken from Arabic, the word in that language being " El Karoub " or as some spell it " El Keurroub."

the thickest of them he rode," leaving his stores well-dressed with arsenic, so that, like Samson in his fall, he killed more of his enemies than in his life. Today he is said to live in Fez, greatly respected, a quiet old Arab with a fine white-beard, whose greatest pleasure is to tell his rosary.

Curious how little the Oriental face is altered by the storms of life. I knew one, Haj Mohammed el ——, —a scoundrel of the deepest dye—who in his youth had poisoned many people, had tortured others, assassinated several with his own hand, and yet was a kindly, courteous, venerable gentleman, whose hobby was to buy any eligible young girl he heard of, to stock his harem. One day I ventured to remark that he was getting rather well on in years to think of such commodities. He answered; " Yes, but then I buy them as you Christians buy pictures—to adorn my house; by Allah, my heirs will be the gainers by my mania." Yet the man's face was quiet and serene, his eyes bright as a sailor's, his countenance as little marred by wrinkles as those one sees upon the Bishops' benches in the House of Lords; and as he stroked his beard, and told his beads, he seemed to me a patriarch after the type of those depicted in the Old Testament. Perhaps it is the lack of railways, with their clatter, smoke, and levelling of all mankind to the most common multiple; but still it is the case—an Eastern scoun-drel's face is finer far than that a Nonconformist Cabinet Minister displays, all spoiled with lines, with puckers round the mouth, a face in which you see all natural passion stultified, and greed and piety—the two most potent factors in his life—writ large and manifest.

An orange grove, backed by a cane-brake, with the canes fluttering like flags, was near to us; cows, goats, and camels roamed about the outskirts of the town, as in Arcadia—that is, of course, the Arcadia of our dreams—or of Theocritus.

Jews went and came, saluting every one, and being answered: "May Allah let you finish out your miserable life"; but yet as pleased as if they had been blessed. Their daughters came, like Rebecca, to the well—all carrying jars —unveiled, and yet secure, for in this land few Moors cast eyes upon the daughter of a Jew.

Upon the ramparts, shadowy white-robed figures, with long guns, go to and fro, guarding the town from hypothetic enemies. Through an arch, between two palm-trees, the Kutubieh rises, distant and slender, and the white haze around its base shows where Marakesh lies. On every hedge are blackberries and travellers' joy; whilst a large honey-suckle, in full flower, smells better than all Bucklersbury in simple time; a jay's harsh cry sounds like the howl of a coyote, and Europe seems a million miles away. In the evening light, the footpaths, which cut every hill, shine out as they had all been painted by some clever artist, who had diluted violet with gold.

Nothing reminds one that within half a mile a town, in which three thousand people live, is near, except the foot-paths which zigzag in and out, crossing the fields, emerging

out of woods, and intersecting one another, as the rails
seem to do at Clapham Junction. A fusillade tells that a
Sherif from the Sus has just arrived; all Amsmiz sally forth
to meet him mounted upon their best, charging right up to
the Sherif, and firing close to his horse—wheeling and yelling like Comanches—making a picture shadowy, fantastic,
and unbelievable to one who but three weeks ago had left
a land of greys. The Sherif's people open fire, and when the
smoke clears off the Sherif himself, mounted upon a fine
white horse, rides slowly forward, holding a green flag. The
people fall into a long line behind him, which, by degrees,
is swallowed up under the horse-shoe gateway of the town.
Mohammed el Hosein, whom I had sent to look out for a
horse, appears with a little, undersized, and brachycephalic
Shillah, leading a young, black horse, unshod, with feet as
long as coops, in good condition and sound in wind and
limb, and with an eye the blackest I have seen in mortal
horse—a sign of perfect temper; incontinently I determine
in my mind to buy him at all cost. We praise our horses—
that is, the Shillah praises his—but I, as a Sherif, am far too
grand to do so at first hand; so Swani lies like a plumber
about mine—which stands and kicks at flies, and looks—
like horses do when one is just about to part with them
—much more attractive than one ever thought before.

The Shillah leads up his horse, which I pretend hardly to
notice; and, luckily for me, he speaks no Arabic. Then he
looks at my horse, and, I imagine, grabs him, for Swani
springs upon the horse's back, and runs him *ventre à terre*

over the rough stubble, and, stopping with a jerk—the horse's feet cutting the ground like skates—gets off, and says that, help him, Allah, the horse is fit for Lord Mohammed, and that I only sell him because I am so tenderhearted about the sore upon his back.

This makes the Shillah think there is a "cat shut up inside,"* for no one sells a horse for a sore back amongst his tribe. We try his horse—that is, Mohammed el Hosein makes him career about the field without a bridle; and then I mount him, and display what I consider horsemanship. Nine dollars and the horse? By Allah, seven; but the Shillah, who has not tried my horse, looks at his colt, and says: " I brought him up, fed him with camel's milk, rode him to war, and he is six years old; nine dollars and the horse." We shake our heads; and he, mounting his horse, yells like a Pampa Indian, and charging through the cane-brake bare-backed, and with nothing but a string on one side of his horse's neck, rushes up a steep bank and disappears, riding like a Numidian. I say it was a pity I did not give the nine, but am assured all is not over, and that the man has probably gone to bring back " a bargain-striker " to complete the sale. Throughout Morocco, when animals change hands, the bargaining lasts sometimes for a week; and at the last a man appears, sometimes a passer-by, who is pressed into the undertaking, who, seizing on the bargainers' right hands, drags them together, and completes the deal.

* " Aqui hay gato encerrado," is the Spanish proverb in reference to anything which seems too good to be true.

In about half-an-hour our man comes back, bringing the local "Maalem," that is smith, who talks and talks, and as he talks surveys me from the corner of his eye. At last the Shillah yields, takes the seven dollars, counts them with great attention, tapping each one of them upon a stone to see if "it speaks true," and then mounting my horse essays its paces. As it fell out the horse, which galloped like a roe, with Swani, set off with a plunge which almost sent the Shillah over one side, and turning flew to the mules and stopping by them refused to move a step. The Shillah thought, of course, he had been done, and I had most reluctantly to mount the horse and make him go. We give our word the horse is not a jibber, and swear by Allah if he is, and the Shillah will send him back to Mogador, he shall receive his money and his own horse on our return.

He takes our word at once and grasps my hand, puts his own bridle on my horse, and bending down kisses his own horse underneath the neck, and says, in Shillah, that he hopes I shall never ill-treat it, I promise (and perform) and bid my own horse farewell after my fashion, and the Shillah mounts and rides out of my life towards the town. A little dour and fish-eyed, turbanless and ragged man, legs bowed from early riding, face marked with scars, a long, curved knife stuck through a greasy belt, hands on a horse as if they came from heaven; his farewell to his horse was much more real than is the leavetaking of most men from their wives, and moved me to the point of being about to call him back and break the bargain, had I not reflected that the

pang once over, the poor Shillah would never in his life be
at the head of so much capital.*

In less than half-an-hour the Maalem had shod the horse,
shortening its feet with an iron instrument shaped like a
trowel, and nailing on the shoes, which almost cover the
whole foot, with home-made nails; he stays to guard our
animals (and spy upon us), and we prepare for our first
al fresco night.

Unlike America, where travellers sleep out of doors from
Winnipeg to Patagonia, in the East, except in crossing
deserts, to sleep out of doors without a tent is quite excep-
tional, and yet one never sleeps so soundly as on a fine night
beside a fire, one's head upon a saddle, feet to the fire, and
the stars to serve as clock. Even the wandering Arabs generally carry tents, and thus, in my opinion, all through the
East much of the charm of camping out is lost. All that we
do is a convention, and Arabs are not savages, but on the
contrary even the Bedouins are highly civilized after their
fashion, and the civilized man must always have a roof,
even of canvas, over his head to shut out nature. Not but
that your common Moors, like to "your Kerne of Ireland"
in ancient days, do not sleep out, for nothing is more common than on a rainy night, in camping at a village, for the
Sheikh to send a guard to watch a traveller's horses, and for
the guard to "liggen in their hoods" all the night long.
So at Amsmiz, under a fig tree, I made my camp, despite

* The Shillah's black horse is now in the hands of Don José Miravent the
Spanish Consul at Mogador, after having carried me all the journey. As to my
grey horse I cannot say, nor yet be certain if there are birds in last year's nests.

the protests of Lutaif, the Maalem,* and Mohammed el
Hosein, who joined in saying that it was not decent for a
man of my (Moorish) position to camp outside the town.
Swani himself was rather nervous, and, as it turned out,
there may have been some risk, for a strong "war-party,"
as they would say upon the frontiers in America, drove
almost every head of cattle belonging to the town that very
night. We slept as sound as dormice, and the Maalem,
who kept watch with an old muzzle-loading rifle stamped
with the Tower mark, slept like a top, knowing the duty of
a most ancient guard, for during the night I thought I
heard a noise of people passing, and waking saw him fast
asleep with the old rifle by his side full cocked, and with a
bunch of rushes in its muzzle either for safety or for some
other reason not made plain. Darwin relates the peculiar
and ineffaceable impression a night he slept under the stars,
upon the Rio Colorado, in Patagonia, made on him. He says
the cold blue-looking sky, the stars, the silence, the dogs
keeping watch, the horses feeding tied to their picket pins,
and the sense of being cut off from all mankind, appealed
to him more than the beauty of the tropics, the grandeur
of the Andes, or anything that he remembered in his travels.
And he was right. Nothing appeals to civilized and to uncivilized alike so much as a fine night when one sleeps near
one's horse, and wakes occasionally to listen to the noises
of the night. Men from the counter, from the university,

---

* "Maalem" literally means a "master" as a master carpenter, master smith,
etc. In Morocco it is often used for a good rider who is said to be "Uahed
Maalem."

riff-raff of towns cast out like dross into a frontier territory, all feel the spell. The Indians and the Arabs feel it, but do not know exactly what they feel; still, in a house, under a roof, they pine for something which I am certain is the open air at night.

'Tis said some ancient wise philosopher once took an Indian from the southern Pampa and showed him all the delights, the pomps and vanities, of Buenos Ayres: showed him the theatre with Christian girls dancing half naked, took him to Mass at the cathedral, led him along the docks, let him fare sumptuously, and then accompanied him to gaze upon the multifarious faces of strange women, who used to lean from almost every balcony and beckon to the stranger in that town; and then the Indian, fed, instructed, with his mind enlarged by all the pomp and circumstance of a Christian town, was asked to say if he liked Buenos Ayres or the Pampa best. The story goes that, after pondering a while, the Indian answered: " Buenos Ayres very large, beautiful things, very wonderful, Christian women kind to poor Indian, but the Pampa best." A brutish answer of a brute mind. 'Tis patent that the man was quite incapable of understanding all he saw; no doubt our gospel truths were all unknown to him, the philosopher who took him round to haunts of low debauchery either a fool or knave; but on the other hand, riddle me this: how many men of cultivation, education, and the rest have seen the Pampa, prairie, desert, or the steppes, and putting off the shackles of their bringing-up, stayed there for life, and become Indians, Arabs, Cossacks, or Gauchos; but who ever saw an Indian, Arab, or wild man of any race come of his own accord and put his neck into the noose of a sedentary life, and end his days a clerk?

And so of faiths. The missionary for all his preaching would never mark a sheep, had he but gospel truths alone to draw upon. What brings the savage to the fold is interest, guns, cotton cloth, rum, tea, sugar, coffee, and the thousand things for which a commentator might search the Scriptures through from end to end and not find mentioned. In Central Africa the Christian and the Moslem missionary are both at work marking their sheep as fast as may be, and each one as much convinced as is the other of the justice of his cause. With a fair field, without the adventitious aids of Christian goods, the Moslem wins hands down.

The Christian comes and says " My negro friends, believe in Christ," and the poor negro, always eager to believe in anything, assents, and then the Christian sets him to black his boots, and all that negro's life he never rises to an equality with his converter. Then comes the Mussulman and cries " Only one God." The fetish worshipper who has had a dozen all his life, thinks it a little hard to give them up, but does so, becomes a Moslem, and is eligible to be Sultan, Basha, Vizier, Kaid, Sheikh, or what not; so that, put rum and rifles on one side, let preaching be the test, in fifty years from the Lake Chad to Cape Town there would not be a single negro, except a few who stuck to their old gods, outside Islam.

I slept but fitfully, knowing if I could leave Amsmiz without suspicion that I was sure to reach my journey's end. The night wore on, the sapphire sky changing to steely-blue; as the muezzin called at the Feyer,* great drops of dew hung from the leaves; up from the river rose the metallic croaking of the frogs; Canopus was just setting; across the sky, the mirage of the dawn stole stealthily, the mules stood hanging down their heads upon the picket rope, a jackal howled somewhere far out upon the plain; Allah perhaps looked down, and not far off from me baited quite peacefully my (new) destrere on " herbes fine and good."

* " El Feyer " in Morocco is the call to prayers about three in the morning.

Chapter V

PLEASING to wake up under the fig trees with all our cloaks and blankets wet with dew, to find our guard, the Maalem, still sound asleep, and be accosted by a tracker who informed us that many of the cattle of the town had been driven off the night before, and who would hardly believe that all our animals were safe. Soon parties from the town, armed and muffled up in cloaks against the morning air, rode out from underneath the horse-shoe gateway, and spread in all directions, trying to strike the trail of the "mad herdsmen," as they were called along the Highland Border in the days before prosperity had rendered Scotland the home of commonplace.

Though sharp of eye, it did not strike me that the Arabs and Shillah were experienced trackers. They rode about too much, and must have crossed the trail of the lost cattle a hundred times, and I kept thinking that an " Arribeño " *
from the River Plate would have done better work than the whole tribe. Still, as they scattered on the hills, their white clothes just appearing now and then behind the clumps of trees, as they quartered all the ground, they fell effectively

* An " Arribeño " is a man from the upper provinces, some of which have long hard names, as Catamarca Jujuy Rioja, etc.; so, to save adjectives, they are lumped as "Arribeños."

into the middle distance, and no doubt if they never found
their cattle, on the first fine night they would recoup themselves from some neighbouring village in the hills. Cattle
and horse-stealing, with an occasional higher flight into the
regions of abduction of young girls, seem to be staple industries in all pastoral countries, and nowhere but in western
Texas are taken very seriously; but there the horse-thief
hangs. Camped once outside a village called Bel-a-rosis, on
the road from Tangier to Rabat, a raiding party attacked
the place upon a rainy night, fought quite a lively action,
the bullets coming through my tent, drove off some horses
and several yoke of oxen, and in the morning the Kaid rode
up on horseback, his followers behind him, in just the spirit
that a gentleman at home starts out to hunt. He praised his
God for his good luck, and said he reckoned (by Allah's
help) to retake all the " Creagh," and drive above a thousand dollars' worth of sheep, of cattle, and horses, from the
weakest neighbours of the nearest hostile tribe he met.

In all the actions that take place consequent upon these
cattle raids, but few are slain, for though both the Arabs and
the Berbers all have guns, their style of fighting is a survival
from the time they carried bows. They rarely charge, and
never engage hand-to-hand, but gallop to and fro and fire
their guns with both eyes shut, or, turning on their horses,
fire over the tail; so few are killed, except the prisoners, who
generally are butchered in cold blood, if they are not of such
account as to be worth a ransom. A merry, pleasant life
enough for able-bodied men, and not unhealthy, and one

that makes them singularly bad subjects for missionary work. So stony is the ground in the vineyard of the tribes, that up to now I never heard of any missionary so bold or foolish as to attempt to dig. A day will come, no doubt, when their hearts will prove more malleable; but I fear before that time their bodies will have to be much wrought upon by rifles, revolvers, and the other civilizing agents which commonly precede the introduction of our faith.

Leaving Amsmiz, the road to Sus leads over foothills all of red argillaceous earth and fissured deeply here and there by winter rains. Now and then strange effects of coloured earth, blue, yellow, green, and mauve, diversify the scene. The road leads through a straggling oak wood and emerges at a village, where, in the middle, the county council is assembled in a thatched mosque. Mosques serve for council chambers, meeting places, and in villages for travellers to sleep in, for throughout Morocco the sanctuary is never closed, and thus the people feel their God is always there, and not laid up in lavender for six days in the week.

The Maalem who had accompanied us as guide, and as a guarantee that we were creditable folk—for in the wilder districts of Morocco travellers take a man in every district to convoy them to the next—had reached his limit, and, laying down his gun, entered the mosque to get another man. What he said there I do not know; if he had doubts perhaps he uttered them, at any rate, after a most unseasonable wait, which kept us all on pins and needles, he emerged, bringing a most ill-favoured tribesman, who came up to me and, kissing my clothes, asked for my blessing. I gave it him as well as I was able, but fancied he was not satisfied, as he retired muttering in Shillah, of which I did not understand a word. The Maalem bade us good-bye, and asked for cartridges, which, of course, we gave, but how they benefited him is a moot point, for they were intended for a small-sized Winchester; he drew his charge and crammed the four or five cartridges we gave him into his old Tower gun, but as I did not pass by Amsmiz on my return, I have no information how he fared when he touched off the piece.

Little by little the road got worse until we entered a tremendous gorge, just like a staircase, which made the worst roads in the Sierra Morena look like Piccadilly by comparison. Only a mule, an Iceland pony, or a horse bred in the mountain districts of Morroco could have coped with such a road, and, as it was, the efforts of the poor brutes were pitiable to see. Under our feet, at a great depth below, the Wad el N'fiss boiled furiously amongst the stones, winding and rewinding like a watch spring, and forcing us to cross it many times, when its swift current proved so formidable that, although not more than three feet deep, we had to enter altogether in a group to keep our feet. As we were toiling up a steep incline, a shouting brought our hearts into our mouths, and, looking back, we saw two mountaineers rushing to intercept us, from a neighbouring hill. No time for any consultation, or to do more than cock our miserable guns and sit quite meekly waiting for the worst. On came the mountaineers, bounding from stone to stone till they appeared upon the road and blocked our way. Long guns, curved daggers, and almost naked save for their long woollen shirts, their side locks flying in the wind, they looked most formidable, and poured out at once a volume of guttural Shillah which sounded menacing. Mohammed el Hosein, who with Ali the muleteer spoke Shillah, interpreted to Swani, who, half in Arabic and half in Spanish (which he called Turkish) informed me what they said.

It seemed the desperadoes wanted us to stop until a little boy brought down some Indian corn and milk, for it appeared it was the custom of those hills never to let a well-dressed Moor pass without some little offering. Of course a poor man, or anyone to whom the maize and milk would have been of great service, passed by unnoticed, as in other lands. I graciously assented, and in due course a boy appeared carrying a wooden bowl, from which I drank and passed it to my followers after the usual grace.* The tribesmen would accept no money, but asked me to say a word for them about their taxes, which were not paid, to their liege lord. As might be expected of me I said I would, and should have done so had I known who their lord was or where he lived, or had I chanced to meet the lord in a more seasonable place. Little by little we entered the zone of Arar trees (Callitris Quadrivalvis) about which Hooker says: " This tree has no congener, its nearest ally being a

---

* Bismillah is the Arab " grace before meat." In rendering thanks to the " Great Giver " they say, " El Ham du lillah," " Praise be to God." Poor heathens, what can be the use of their troubling *our* Creator.

South African genus of Cupressus (Widdingtoniana)."* Various writers, as Shaw† and Daubeny,‡ have attempted to identify this tree with the " Θυια," of Theophrastus, and the Thyine Wood of the Apocalypse. Daubeny makes both Martial and Lucian refer to it under the name of Citrus, and Hooker is of opinion that the " Citrea mensa" of Petronius Arbiter was made of Arar-wood. An ingenious writer (Captain Cook, " Sketches in Spain ") is almost certain that the roof of the Mosque of Cordova is of Arar, because the Spaniards refer to it as being made of Alerce. Alerce happens to mean larch in Spanish, but when did a derivation hunter ever allow that the people of any country really understood the meaning of words in their own paltry language ? It is perhaps well not to question too closely the dogmas of science, and what prudent man would aver that anything might not be found in the veracious pages of the Apocalypse?

The Arars spread all over the mountains in a dense low scrub, rarely attaining more than ten feet high, on account of the frequent fires which the inhabitants light in the spring to obtain pasture for their goats and sheep. In an angle of the road I passed four fine old trees, perhaps forty feet in height, immensely branchy, and with trunks measuring possibly two-and-a-half feet in diameter. In the glades between the cedars patches of Indian corn were ripening; some was dead ripe, but no one dared to gather it, for the

* Hooker's " Morocco," page 389.
† " Travels in Barbary."
‡ " Trees and Shrubs of the Ancients."

Kaid of Kintafi, into whose territory we now had passed, had issued orders that he would give the signal for the harvest, and had not yet seen fit to give the word. Well do the Spaniards say that, "fear, not walls protect the vine-yard,"* for not a house stood within several miles of some of the maize patches and still not a head of corn was touched.

The Kaid of Kintafi, who was destined to be my captor, had some years before waged war with Mulai el Hassan, the late Sultan of Morocco. The Sultan had invaded the mountain territory of Kintafi, and destroyed most of the villages, but not being able to reduce the fortress of the Kaid had won him to his side by offering him the hand of either his sister or his daughter, and had departed after having brought the war to a conclusion, on the peace-with-honour plan. The Kaid rendered him henceforth a limited obedience, responding to his overlord's demands for taxes, and for assistance in the field when he thought fit, but taking care not to rebuild his villages, so that the aforesaid overlord might find a desert through which to pass if the fit took him to again commence hostilities. El Kintafi, for after the Scottish fashion he is designated territorially, is of the Berber race, and although speaking Arabic he speaks it as foreigner. He is reckoned one of the greatest chieftains of the Berber people in the south.

Eight or nine months of drought had rendered the plain country through which I passed up to Amsmiz almost a

* Miedo guarda viñas y no vallados.

desert, and it was pitiful to see the people digging for a kind of earth-nut, locally known as Yerna, white in colour and semi-poisonous till after many washings. The plant I have not been able to identify, but the leaves are something like one of the umbelliferæ. Hooker does not seem to notice it in his Botany of Morocco, and perhaps, when he was in the south, famine had not forced the people to dig for it, or the small thin leaves had all been withered by the heat.

Crawling along the mountain roads I found myself trying to estimate which of the two entailed most misery upon mankind, the old-time famine, which I saw going on all round me in Morocco, caused by want of water and failure of the crops, or the artificial modern and economic famine so familiar in·all large towns, where in the West End the rich die from a plethora of food, and in the East End the poor exist just at subsistence limit by continual work. No doubt, in modern towns, the poor enjoy the blessings of improved sanitation, gas, and impure water, laid on in insufficient quantities to every house; of education, that is, illusory instruction to the fifth standard, to fit them to drive carts and tend machines; but, on the other hand, they have but little sun, either external or internal, in their lives, and know their misery by the help of the education which they pay for through the rates.

The sufferer by famine, as in Morocco, suffers enough, God knows, stalks about like a skeleton, dies behind a saint's tomb; but in the sun. He believes in Allah to the last, and dies a man, his eyesight not impaired by watching wheels whirr round to make a sweater rich, his hands not gnarled with useless toil (for what can be more useless than to work all through your life for some one else ?), and his emaciated face still human, and not made gnomish by work, drink, and east wind, like the poor Christian scarecrows of Glasgow, Manchester, and those accursed " solfataras," the Yorkshire manufacturing towns. But place it where you will, either in tents, on some oasis in the desert, wandering as the Indians used to wander before America was but one vast advertisement for pills, or in the sweaty, sooty, noisy "centre of industry," mankind is made to suffer; so, perhaps, Tolstoi has the root of things when he suggests that marrying, giving in marriage, and all licensed or illegitimate · propagation should cease, and man by non-existence at last attain to bliss.

So in a pass between two walls of earth, with bands of shale crossing them transversely, and roots of long dead Arars sticking through the ground, I came upon the human comedy fairly played out by representative marionettes of every age and sex. First, the father, a fine old Arab, gaunt, miserable, grey-headed, ragged, hollow-cheeked, without a turban, shoes, or waist-belt, and carrying a child which looked over his shoulder, with enormous black and starving eyes; the mother on foot, in rags and shoeless, and still holding between her teeth a ragged haik to veil her misery from the passer-by, a baby at her back, and in her hand a branch torn from an olive-tree to switch off flies; then three ophthalmic children, with flies buzzing about their eyelids;

lastly, the eldest son stolidly sitting in despair beside a fallen donkey carrying salt, and rubbed by girth, by crupper, and by pack ropes, and an epitome of the last stage of famine and of overwork. And as we came upon them, from a saint's tomb near by, a quavering call to prayers rang out, and the whole family fell to giving praise to him who sendeth hunger, famine, withholds the rain, and shows his power upon the sons of men, infidel or believer, Turk, Christian, Moor, and Jew, with such impartiality that at times one thinks indeed that he is God. As for the donkey; Bible, Koran, and all the rest of sacred books were writ for man, and the galled jade may wince until the end of time for all Jehovah, Allah, Obi, and the rest of the Olympians seem to care. All our bread, dates, and tea went to the owners of the donkey, and had I been in European clothes I should have bought the starveling beast and put a bullet through his head.

Cistus and heath, with mignonette, dwarf arbutus and stunted algarrobas, with thyme and sweet germander, made a thick underwood upon the hills; and yet, as is the case in all Morocco, and, I think, everywhere throughout the East, footpaths crossed here and there, men seemed to be eternally coming and going, donkeys, more or less wretched in appearance, wandered here and there, and in the air, above the scent of flowers, hung the stench of human excrement, the "bouquet d'Orient," the perfume which, I fancy, scents the breeze in Araby the blest.

Along the desert trails, in the Sahara* and the Soudan,

* Sahara is a dactyl in Arabic.

I fancy, man is rarely long unseen, and camels and don-keys must have been struggling across the sands before the first instant of uselessly recorded time. Still, as in India, wild beasts thrive almost alongside cultivated fields, and in the sandy paths which ran through the thick under-growth tracks of wild boar appeared; whilst, on the bor-ders of the bare hills, above the vegetation, " moufflon " * skipped, looking so like to goats that I could scarcely credit they were wild.

All the streams we crossed had whitish beds, and a white sediment clung to the grass upon their banks. It seemed there was a salt mine in the neighbourhood which supplied all the province, and underneath us far below, looking like ants, we saw long strings of mules and don-keys meandering along the paths towards the mine. The road gradually got worse and worse, and in few places averaged more than four feet wide, so that I rode one stirrup brushing the mountain and in great terror that I should lose my yellow slipper down the precipice, five or six hundred feet in depth upon the other side, which went sheer down into the Wad N'fiss. Occasionally we had to call to trains of mules advancing to stop till we could get into one of the hollows scooped here and there

---

* The " moufflon " of the Atlas is called " Oudad " by the Berbers. No doubt when duly stuffed and labelled in a museum he has his proper Latin name, without which no self-respecting beast can die. People then gaze at him through dusty glass, and the less educated, seeing the Latin ticket, go away wondering at the depth of wisdom men of science seem to descend to.

into the hillside to allow the travellers to pass. The mules
in nearly every instance had their packs covered up in
the striped blankets made in Sus, and woven in a pattern
of alternate black and white bars, with fringed edges, and
curious cabalistic-looking figures in the corner of the web.
Now and again a Sheikh's house, perched upon a hill and
built like a castle, with turrets, battlements, and Almenas,*
all in mud, and looking as like a properly constituted
fortalice as the difference of income of their owners and
that of the owners of the modern stucco fortalice outside
an English town permits. Along the road one constant
interchange of " Peace be with you " was kept up, as we
met other parties of believers going or coming from the
Sus, for the other two main highways being closed by the
outbreak of hostilities, all traffic had converged upon the
road which started from Amsmiz.

Rounding a corner and dipping down a steep incline,
almost before we were aware of it, we found ourselves
between a castle and a small mosque; all round the " per-
ron " which led to the gate groups of muleteers sat resting;
by a side door were lounging several attendants armed
with long guns, the barrels hooped with silver and with
brass, and, to complete the picture the Sheikh himself in

* Casa de Almenas (a house with battlements) is in rural Spain a euphuism for
a gentleman's house. I fancy, like the torch extinguishers in the regions about
Berkeley Square, that these almenas occasionally rise in a single night in houses
where owners in the past have neglected to be legally constituted and known as
gentlemen.

shirt sleeves, so to speak, that is, without a haik, and dressed in a long white garment, sat in the shade studying the Koran. His air of patriarchal simplicity so impressed me that only with an effort I remembered that I too was one of the same faith, and rather rudely I fear, according to Moorish ideas of etiquette, I mumbled a low " Peace be with you " as I passed. From the mosque windows and the door came out a sound of voices as of children shouting all at once; it was the children of the place learning to read, for amongst Arabs no place, however small, no Duar of six tents, lost in some far remote oasis of the Soudan, is without its schoolmaster. About a league farther along the road, and leaving the castle, which rejoiced in the ap- pellation of Taguaydirt-el-Bur, we got off under some ole- ander and olive trees by the side of a clear stream to wash and eat.

Mohammed el Hosein swears by Allah no infidel has ever seen the place, and asks, if so, let him describe it. A thing indeed which Maupassant himself could hardly have achieved without inspection.

" Give me the signs of such a place " is a common Arab question if, in talking of some place, he doubts you have been there. Most of the famous places in the Mohammedan world are well known by description to all believers, and nothing is more common than to hear one man " giving the signs " of Mecca, Medinah, or some other holy spot with the minuteness of a lady novelist describing scenery. As we sat eating under our olive trees, people kept passing on the road continuously. Thus in the heart of one of the most unknown ranges of mountains (to the Nazarene) in the whole world, there is no solitude, no sense of loneliness in spite of all the grandeur of the hills, the snow, the precipices and the brawling streams.

Though I know no single Oriental country but Morocco, and that is known to all Mohammedans as Mogreb-el-Acksa (the Far West), I yet imagine that throughout the East the interest lies entirely in mankind, for nature, at least to judge by what is seen throughout Morocco, is as dominated by man as is the docile soil of England, which gives its crop year in year out, suffers and has endured a thousand years of ploughing, dunging, reaping, draining, and the like, and thinks as little of rebellion as does the mouthy Radical who on his cart thunders against the Queen, and slavers doggishly before a new created lord. Still, for a thousand (perhaps ten thousand) years the Oriental life has altered little, nothing having been done to "improve" the land, as the Americans ingenuously say. And so may Allah please, bicycles, Gatling guns, and all the want of circumstance of modern life not intervening, it may yet endure when the remembrance of our shoddy paradise has fallen into well-merited contempt.

Our local guide, a long, thin, scrofulous-looking Berber, dressed in a single garment like a nightgown, but most intelligent, said that close to where we sat, two hundred years ago lived El Kalsadi, a writer on arithmetic. Casting about the corners of my recollection I recalled having seen the name in Quaritch's Catalogue of Arab books; thus Gentile, Jew, and True Believer for once concurred about one circumstance. I took my saddle cloth and under some oleanders by the stream fell fast asleep, but fifty yards removed away from the men and animals, and, waking, found two Berber tribesmen sitting near to me washing their feet. They turned upon my moving and said something in their own tongue. I, not understanding, put on a grave demeanour and answered, Allah, which seemed to satisfy them as entirely as if I had been able to rejoin as St. Paul advises, with my understanding; but no doubt, the gift of tongues and all duly allowed for, he had often found himself similarly situated when on the tramp. Just at the crossing of this stream we paid and dismissed the guide who had accompanied us from the wayside mosque; he straight departed by a road fit for but partridges alone, across the hills, and I think somehow or other in conjunction with the Maalem from Amsmiz, contrived to pass the word along the road, of their opinion of my adherence to the Christian faith.

Struggling along over the shingly bed of the Wad N'fiss for about two miles, we came again to a funnel-shaped gorge, which led by degrees to a terrific staircase of rocks, known to the local muleteers as N'fad Abu Hamed, the Knees of Father Hamed, as being so steep that no one could ascend it except upon all fours, or from its effects upon the knees of the unlucky mules, or from some other reason as to which commentators have disagreed. Just at the top I looked back over the finest range of mountain scenery I remember to have seen, and thought that with good luck and patriotism some few score Afridis in such a place might hold in check a regiment of marauding Britishers, but on reflection I saw clearly that the word patriotism was out of place when used against ourselves. To the west towards the Sus the mountains seemed to dip, and Ouichidan, the highest peak in all the Southern Atlas, towered right above us a little to the east of where we stood, whilst far away in the dim distance rose the far distant mountains of the anti-Atlas, which rise above the province of the Draa, and on whose snow no Christian has ever trod. Descending rapidly we passed a strath in which some country people were engaged in burning down a tree, an ancient poplar which the wind had partly shifted from its place, and which, when it fell, must fall inevitably across the road. In a small patch of Indian corn an aged negro was at work, dressed all in white with a red fez, and illus-trating (in the middle distance) the mystery of creation in a fantastic fashion of his own.

Stumbling and tripping, leading our exhausted beasts, we came long after nightfall to a miserable house perched on a ledge of rocks over a river, and with a garden be-tween the hillside and the stream; the place is called in Shillah, Imgordim. Nothing was to be had either for love or money, and we went supperless to bed save for green tea, sadly reflecting that had I been a Moor indeed, food would have been obtainable, for I could have sent a man to take it in the usual way, by force; and as it was, perhaps the people though me a fool for not following out the course adopted by all respectable Sherifs when travelling about. Throughout Morocco Sherifs, especially from Fez, are a positive curse to the poorer class of Moors; for being " holy " they behave exactly in the same way that the two Jewish Sherifs Hophni and Phinehas behaved in Holy Writ; but most unfortunately, the people are so unresisting, that punishment which overtook their prototypes seldom is meted out to them, and they pervade the land, begging and taking contributions from the poor, for Allah's sake. One of the really wise laws the French have made in Algeria is that forbidding the Sherifs to go round sorning on the people. Imagine if religion had the same hold in England as it has in Morocco, and recognized descendants of St. Paul* perambulated up and down, taking by force when the faithful were disinclined to give, and you may have an inkling of how onerous the burden was upon the poorer villages before the French put down the practice in Algeria. All the night long the people keep passing on the road, and a small black mosquito, resembling the Jejen of Paraguay, was most impertinent.

Next morning (October 20th) we were early upon the road, as is most generally the case in travel when one goes

---

* I fear though they would have to be descended from St. Peter, who carried a wife about with him, as his great rival somewhat tartly remarks. At any rate, even the Ebioim have never ventured to cast any doubt on St. Paul's private character. This is consolatory when one recalls the case of Burns and other poets, including King David.

supperless to bed and rises to get on horseback with but
a cup of weak green tea for breakfast. The animals had
eaten well as we had barley with us, and, as on the day
before, we started climbing staircases of rock, crossing and
re-crossing the Wad el N'fiss, passing deserted villages, and
interchanging greetings with the various parties of travellers upon the road. A precipitous descent over red sandstone
rocks and through a thicket of tall oleanders, brought us
to an open space, where the bed of the N'fiss spread out
about a quarter of a mile in width. To right and left, at
the distance of a mile or two, on either hand perched upon
hills were fortalices, and farther up the valley a castle
which I was destined to become better acquainted with
before night. The mountains, about ten miles away, formed
a most perfect semi-circle, and Tisi Nemiri,* upon our
right, appeared to rise sheer from the river into the clouds.

Mohammed el Hosein, who, all the morning, after the
fashion of a hungry man, had been exceedingly ill-
humoured, now produced from the recesses of his saddle-
bags four rather small pomegranates, which all five of us
proceeded to devour with great alacrity. Whilst eating, he
informed us that the semi-circle in front was the last ridge
between us and the Sus, and that with luck we ought to
camp well in the province of the Sus by evening.

Again we fell discussing our plan of action on reaching
Tarudant. All now depended upon passing the guard at

* Tisi = hill, in Schluoch. Nemiri means " stones."

the N'Zala,* where an official of the Sultan was reported
to be permanently sitting at the receipt of such custom
as might come his way. Mohammed el Hosein was confident that, for a dollar, he could get us past without much
risk, I shamming ill and sitting a little apart under a tree,
whilst he bargained with the official, the danger being that
the man should, out of civility, put some question to me
which I must answer, and thus, by my poor Arabic, betray
myself, for Mohammed el Hosein was sure that no one
on the road had for a moment questioned my identity.

In fact, about an hour before we camped, whilst riding
along one of the shingly stretches over which the road
went following the bed of the N'fiss, I had been put to
a pretty strong test, and had emerged triumphantly. About
a quarter of a mile in front of us we saw a band of twenty
people sitting on a grassy slope and watching apparently
for our passing by. Of course we thought it was a detach-
ment of tribesmen who had penetrated my disguise and
were about to attack us, but, as to turn back was quite
impossible, and even consultation difficult, by reason of a
party of travellers who followed us, we had to face the
difficulty, after having deliberated briefly upon my drop-
ping my handkerchief to give Mohammed el Hosein the
opportunity to get off and ride beside me to consult. As
we came closer I saw the people had no arms and were
chiefly young men of from eighteen to twenty years of

* N'Zala is a sort of post-house established by the Government. At the N'Zala
we contemplated passing there was a sort of custom-house put up to swindle
travellers, as is usual in all well-governed countries.

age. Swani passed on the word it was a Tolba, that is, a
band of wandering students, in fact, an Estudiantina, and
then I knew what line of conduct to adopt. Just as we
crossed the river they came all round my horse, wading
in the fierce current above knee deep, and, catching my
cloak, kissed it most fervently, invoking blessings on the
Sherif from Fez, and calling upon Sidi bel Abbas, the
patron saint of travellers, to guard me from all harm. I
knew of course that that meant money, and blessed them
in my best Arabic, whilst luckily for me the splashing of
the horses and the men crossing the stream made so much
noise I might have spoken in Chinook for all they knew.
I signed to Swani who advanced and gave them a peseta,
and some of them kissed my knees, which I permitted
graciously as to the manner born, and, stooping down,
touched with my right hand the head of one who seemed
the cleanest, and rode on, looking as ineffable as a man
desperately anxious and engaged in crossing a shingly,
violent-running river can be supposed to look. Mohammed
el Hosein and Swani, the danger past, and the travellers behind us having halted, laughed till they almost fell off
their pack saddles; Lutaif, I think, gave thanks after the
fashion of the Lebanon, and Ali looked at me, thinking,
I verily believe, I really was as holy as I hope I looked.

Naturally, this little passage raised our spirits, as the
students of the Tolba, though beggars, were all educated
men, had wandered much about the country, and must
have seen Christians at the coast towns a thousand times.

I recommend to educational reformers a peripatetic university after the Tolba system. The students might beg their way along on bicycles, listen to lectures in convenient places —upon commons, under railway arches, beside canal locks, and the like—play football, as the members of the Tolba in Morocco play, by the roadsides, and thus high thinking with low diet, football, and bicycle, might all conduce to scholarship, to health, and knowledge of the country roads. In Morocco the system works well enough, although some say that in the Tolbas those customs flourish which have led scholars in England to Malebolge. Even of stationary universities ill-natured things have been preferred. Be all that as it may, it seems a merry life to lounge along, to read under the olive trees, camp near some village, write charms for folk who want them, and recite whole chapters from the Koran, at night, yelling them out in chorus, after the way of children learning their lessons in a kindergarten.

So, thinking on the Tolba, our spirits rose, and we determined to send Ali to the nearest house to buy provisions. Lucky we did so, for it provided us with a square meal which we were destined not to enjoy again for several days. It appeared, the N'Zala once passed, that a sharp descent of some three hours led into the plain beyond. There, according to Mohammed el Hosein, the people were all armed and very warlike; and the road to Tarudant led through the main streets of several villages. He did not look forward to much risk in passing them, but thought there might be a chance of falling in with robbers by the

way, who naturally would rob a Moor (they being Berbers) quite cheerfully, and would without doubt kill any
Christian whom they came across, not being restrained
by fear of the authorities, as are the people in the settled
parts of the country across the mountain. However, Mohammed el Hosein finished up always by saying, "We
are in Allah's hand; but leave it all to me, for if danger
should occur it is not for nothing I am called the cleverest
muleteer upon the southern road." And as he spoke his
little Shillah eyes sparkled like live coals, his thin, black,
pointed beard wagged to and fro, and his face and muscular
arms twitched and contracted as he shook with laughter
in the enjoyment of the joke. To deceive anyone is always
pleasing to a Moor (sometimes to Christians also), and
to take in a town and pass an infidel upon it as a Sherif
from Fez appeared to him the greatest piece of humour
of his life. To our enquiries as to what was best to do if,
in a village, I was recognized, his answer was invariably
the same, so that at last I did not bother him, seeing him
confident, and feeling almost certain in my mind the worst
was past. Ali returned laden with bread and mutton, eggs
and fruit, and we sat down to eat our last repast in freedom
in the jurisdiction of the Kaid of Kintafi, whom Ali told
us lived like a Sultan, and that he had had to wait full
half an hour with other travellers before being permitted to
purchase food.

Lunch over, we got on the road in great good humour,
and for an hour crossed a bare stony plain, till, winding

round a little hill, we came suddenly into full view of a deserted house on one side of the road, and about half-a-mile away upon the right an immense castle surrounded by gardens, woods, and cultivated grounds, and with the river el N'fiss flowing just underneath the walls. Mohammed el Hosein knew the place well, and said it was called Talet el Jacub, the summer residence of the Kaid el Kintafi, the governor of the province that we were journeying through. "Please God, he is not on the look-out for us," he said half laughing; and as he spoke a messenger came running to meet us, his clothes tucked into his belt, bare-headed, and a long staff in one hand. We took no notice, and he overtook us and asked where we were going, and Mohammed el Hosein replied, "Towards God's land," an Arabic retort to an inquisitive enquiry on the road. The messenger retorted, "This is no laughing matter, a man came to the Kaid's house this morning and said he had heard there was a Christian on the road disguised as a Mohammedan." Luckily all this passed in Shillah, and the speaker scarcely knew as much Arabic as I myself. I called up Swani and told him to tell Mohammed el Hosein that I was going to see Basha Hamou, at Tarudant, and that I had not time to call upon the Kaid, as I intended to camp that night in Sus. The man looked at me, at Swani, and at Lutaif, who spoke to him in Arabic, and he said, "Then you are not Christians?" to which Swani replied, "No, burn their fathers;" and the messenger, after profuse apologies, returned towards the castle at a dog-trot.

Who now so certain as ourselves of our arrival at Tarudant? We agreed the Kaid will stop all passers-by and lose much time, and in five hours at most we shall be past his jurisdiction, and it is not in the Sus that he will follow us, even should he discover his mistake.

So we spurred on quite merrily, laughing and talking of the rage the Kaid would fall into when he heard some day how near he had had the Christians in his hand. Past walnut woods, through thickets of scrub oak, by gardens into which the water ran through trunks of hollow trees, upwards steeply ran the road, passing by hedges thick with brambles and dog roses, giving a look of Spain or Portugal, and every step we went we laughed at the discomfiture of the foolish Kaid.

After an hour of steep ascent over the shoulder of a mountain called Tisi in Test (Hill of the Oaks), we struck a steep staircase of rocks, and Mohammed el Hosein said, " In an hour from here we shall pass a castle by the roadside, it is the guard-house of the Kaid, and from thence to the N'Zala is but half an hour. Once there, in a few hours you will see the tall towers of the mosques in Tarudant." So we determined (it was then about one o'clock) to push on without eating and sleep in Sus. The steep ascent proved steeper than any we had passed, but we cared nothing for it, knowing we were so near our goal.

At last we neared the castle by the roadside, no one seemed stirring near it, and we were just about to pass the gate when a loud shouting just below us made us turn our heads. To our amazement we saw our friend the messenger accompanied by several well-armed men, bounding up the steep road like an Oudad (moufflon), and shouting, in Shillah, in a voice to wake the dead. Men rushed out of the castle and ran for their horses, and the messenger arrived just as we were about to pass the door. We stopped, and putting on an air of quiet citizens, alarmed upon the road, asked what the matter was, although we knew. Men rushed and seized our animals, called out "Arrumin!" that is, "The Christians!" brandished their guns, fingered their daggers, and for a moment things looked ugly. I sat upon my horse hardly quite catching all that was said. Lutaif expostulated and Swani, calling on Allah, asked the Sheikh, who now had come out of his house, and stood waiting till some one brought him his horse, if he looked like a Christian. "No," said the Sheikh, "you appear to be a cursed sailor from the coast, accustomed to sail upon the black water, and to consort no doubt with Christians." Swani looked as if he would have liked some private conversation with the Sheikh near Tangier, but prudently said nothing, and the Sheikh turned to Lutaif and asked him who he was. Lutaif replied, "A Syrian and a Taleb, and the attendant of this gentleman," pointing to me.

"Then," said the Sheikh, "this is the Rumi," and, turning to me, said, "Is it not so, or will you swear you are a true believer?" Swearing is easy if you possess a language pretty well, but difficult in "petit negre," and so, knowing we should be taken back before the Kaid and then found out, I answered "Yes, I am the Christian," and began to feel my horse's mouth ready for what might come.

As Allah willed it nothing occurred beyond a little shouting, and some rather tempestuous brandishing of guns, and threatening looks. The Sheikh, who by this time had got upon his horse, rode up to me and looked me in the face. I said, "Have you ever seen a European before?" but his Arabic was at an end, and the rest passed in Shillah between Mohammed el Hosein and the man sent by the Governor.

It appeared the messenger had gone back to the castle and told the Kaid we were not Christians, and that I seemed a reputable man, riding a good black horse. On that the Kaid exclaimed, "Black horse, I was told the Christian bought a black horse in Amsmiz, so after them at once with four or five armed men"; and to teach him circumspection and lightness of foot, had him well beaten before he sent him off. A guard of men advanced and took our bridles and began to lead us back, as downcast a company as you might see upon a long day's march. I felt like Perkin Warbeck going to the Tower, and rode quite silent, but cursing under my breath, whilst, as I take it, the loud jokes which passed in Shillah were most amusing to our captors, though I feel doubtful, even had I understood them, that they would have amused me in the least. One ragged tribesman tried to snatch my gun, which I had borrowed from the Consul in Mogador, but I kept hold of it and told him that if it no longer belonged to me it was the Kaid's, and I should tell him of the theft.

Nothing takes better with the Arabs and Berbers than an answer of this kind, and when he understood it the man grinned like a baboon and said he was no thief. I had my own opinion about this, but thought it wise not to disclose it, and at that moment a heavy storm of rain swept down from off the hills and wet us to the skin. We now began to press our animals, and the escort bounded like goats beside us, one of them trying to prick my horse with a long knife to make him kick. The headman, seeing the joke, promptly struck him across the head with a thick stick, the blow being enough to have stunned a European, but which did not seem much to annoy him and he trotted along just like a hound who has received a cut from the second whip for running a false scent.

In about three-quarters of an hour we did that which had taken us two hours to come, our animals rushing down the steep paths in the heavy storm, and the escort shouting and cursing like demoniacs. We plunged into a wood, crossed a flooded stream, rode through a field of standing corn, and, crossing the maidan* before the castle, came to a horse-shoe arch. Assembled before the entrance was a crowd of armed retainers, loafers, herdsmen, travellers, and all the riff-raff who, in Morocco, haunt the dwellings of rich men. Boys, and more boys, oxen, and goats, and horses, all pressed into the gateway and the dark winding passage, to escape the storm. Loud rose the cry of "Christians, sons

---

* A "maidan" is an open space on which to practise horsemanship, and one is generally to be found before the gate or near to the walls of every castle or Sheikh's house throughout Morocco.

of dogs." I thought, in the dark passage, that the occasion seemed quite favourable for some believer to strike a quiet blow for Allah's sake.

Swani pushed forward, and placed himself beside me on his mule; so pushing, striving, cursing, and dripping with the rain, we reached a second gateway, which opened into a court; and here the travellers, herdsmen, and the rest entered, and left us, with our escort, standing in semi-darkness below the arch. For a full hour we waited—I, sitting on my horse, partly from pride and partly from the instinctive feeling of a horseman that he is always safer on his horse; the others dismounted, and sat down on a stucco bench, looking the picture of misery and discontent. We did not talk much—though I felt inclined to laugh— the position striking me as comic enough in all its aspects; and at last a fat man, in immaculate white robes, holding a bunch of keys, came through the rain across the inner court, and asked my name. I told it to him, and he seemed edified, and asked what I was doing, and why I wished to go to Tarudant. This was most difficult to explain, for in Morocco few people journey out of curiosity, so I replied I had business with the Governor. The man then turned towards Lutaif, and said: " You are a Mohammedan, why do you travel with this Christian? You are a taleb, and should know better than to connive at a Christian travel-ling in Moorish dress." Lutaif had to explain he was a Christian, and the fat man then turned upon Mohammed el Hosein, and said: " You are no Christian, and the Kaid

says he will shave your beard, give you the stick, and put
you into prison with some comfortable irons on your feet."
To such a speech there was no very obvious answer but
" Allah Kerim," which poor Mohammed el Hosein mumbled piteously enough; and the fat man, kilting his snow-
white robes, waddled across the court, and went into the
house, without another word.

And so we waited, as it seemed hours, till again across
the court came the fat, snow-white-robed official, accompanied by a short, broad-shouldered man with a full black
beard, who, walking up to me, held out a hand, and said,
" Bon jour." I answered, but his French extended no
further, and he tried Turkish—of which I did not know
a word. Lutaif, who spoke it well, entered into a conversation with him, when it appeared he was a Persian—a
sort of wandering minstrel—who was staying with the
Kaid and had been sent, upon the strength of his " Bon
jour," to find out who we were. As I was still uncertain
if the Globe Venture Syndicate's steamer was off the coast,
we took good care to make it plain we were friends of
Basha Hamou, for, had the vessel arrived, we should have
been thrown into prison at once, and sent in chains up to
the Sultan's camp. However, no suspicion of this seemed
to cross the people's minds, and we sat on talking to the
old Persian in a jargon of mixed Arabic and Turkish for
some time, whilst gradually a crowd of people had assembled, who—sitting on the ground, on stones, and on
a low divan which ran right round the arches—glared at

us silently, like men looking at wild beasts. A boy or two threw a few stones, but they were stopped immediately—the order having evidently gone out to treat us well. Considering all things, and how completely we were in their power, how far removed from any Europeans, and how strong the spirit of dislike is to all strangers, especially amongst the Berber tribes, the conduct of the people was quite wonderful; and I question very much if in a European country we should, in similar conditions, have fared as well. Still, the uncertainty made waiting anxious work, and we were pleased to hear from the old Persian that the Kaid was in a dilemma as to what he should do with us—whether to send us back, let us go on, or write for instructions to the Sultan as to what course to take. At last the Chamberlain, wrapped this time in one of the Atlas brown goat-hair cloaks—called by the Berber an Achnif—came back again, and said the Kaid had made his mind up to extend his hospitality to us, and that he had placed a tent at our disposal on the Maidan. The phrase admitted no discussion; and so following the Chamberlain—and preceded by an attendant with a long gun—we rode to the Maidan, and found a tent pitched on the wet ground above the bank of the N'fiss, into which all of us, with saddles and baggage, were glad to pack, to get out of the rain. The tent was large, and circular in shape, ornamented outside with rows of blue cloth decanters—after the Moorish fashion—and lined with a chintz of the most pre-Morris kind. In addition to its beauties of form and decoration it leaked in several places, and was so loosely pitched that we had to turn to at once to make it safe, and dig a trench to carry off the water, which stood about an inch in depth upon the floor. Five men with pack-saddles, bridles, and guns, and all the requisites for camping, left very little room to move about. A forlorn crew we must have looked as we sat, shivering and hungry, on the damp floor. Closing the door, I drew from my saddle-bags a bottle of brandy which I had in reserve for snake-bites, and administering a dose (medicinally) to believers and infidels alike, a better spirit soon prevailed, and we got beds made down—placing some stones to keep the blankets off the ground. A fire was lighted; and as we drank our tea—flavoured with some mint which Ali drew out of his bag, where it had lain for days amongst tobacco, pieces of string, and the " menavellings " of a muleteer's profession—we set about reviewing our position, after a joke or two, as to the enjoyment of the hospitality so generously provided by the Kaid. For myself, I was not in much trepidation, knowing the worst that could occur was to be sent under a guard to the Sultan's camp—a matter of from five to six days' journey. As to Mohammed el Hosein, that was more serious. Already he had been threatened with the stick, imprisonment, and with the loss of beard—the greatest insult which can be put on a Mohammedan. But the poor devil (and we ourselves) knew well that in his case the stick most probably meant death, and that he would not live to undergo the other punishment. Still, he was not so much downcast as might have been expected, but sat in the wet mud—a bellows in his hand—blowing the charcoal for the tea; and said, resignedly, "We are in Allah's hand; but it is a pity I was but newly married before leaving Mogador."

Swani, as a man, so to speak, without caste, was safe, and the most he could expect was a few dozen blows with a stout stick, a matter about which he did not seem to care a halfpenny, for he knew that he could plead he came under compulsion as my servant, and the plea would almost certainly be held sufficient at least to save his life.

Ali, of course, could not be held an accessory before the fact, as he left Mogador without the least idea of our intention, and I assured him if he lost his mule that I would buy him one fit for the Sultan's saddle.

As we sat talking, we perceived that a group of tribesmen, all fully armed, had sat down just outside the tent, the rain having ceased for a little, and were regarding us quite motionless, but with their eyes not losing any action that we made. Of course, we were the strangest spectacle they had ever seen, and after half an hour, their curiosity well satisfied, they moved off silently and sat down in the same manner to watch a game of football, which was proceeding on the Maidan, and in which all the young men, from slaves to the Kaid's sons, were taking part. So we resumed deliberations, and discussed the position of Lutaif. He was perhaps, of all of us, in the most dangerous case. A Syrian and a Turkish subject, without a paper of pro-

tection from any European Consul, as he said himself:
"If the Kaid wants to kill me, he will do so as if I were
a dog, and you may be certain that my Sultan will not
claim compensation for the death of any Christian." I
thought about Armenia, but the time was scarcely oppor-
tune for joking; and just as all our spirits and our stomachs
were at the lowest ebb, a slave came from the Kaid, bringing a dish of couscousou,* which we devoured at once,
and could have eaten at least five times as much. The
canvas door was lifted, and, with a cheerful but irrelevant
Bon jour the Persian entered, sat down without a word
and, after looking at me for a moment or two, said "Mez-
quin" (that is, "poor fellow"), "how far you are from
home." This, though a truism, had not occurred to me,
but put thus, à brûle pourpoint, it seemed to come home
with great force to Lutaif, to Swani, and to Mohammed
el Hosein, and for a moment they seemed about to weep,
after the fashion of the two aged men who wept because
they both were orphans.

The Persian promised "to stand before the Kaid and
speak for us," and to return tomorrow and relate his life
to us. We all bid him good night with great effusion, as
he had been a valued friend, and watched him walk across
the Maidan into the castle, and perceived that round our

---

* Couscousou, or cuscus as it is often called in Morocco, is something like the
American hominy. It is made of wheat pounded and grated, and then steamed,
so that the result is a very dry porridge. You eat it with your hands, and the result
is rather messy to the novice. Sometimes mutton and pieces of pumpkin are
served up on the top of the cuscus, and on grand occasions it is made with sugar
and milk, and flavoured with cinnamon.

tent, some fifty yards away, on every side, squatted a sentinel.

Whilst we sat trying to dry our clothes, from the castle mosque broke out the call to prayers, called by the Persian in a voice like the last trumpet's sound, the tower seemed to rock, and the hills gave back God's name from every crag and hollow, till the whole valley quivered with the sound, and the night air was all pervaded by the echoing cry. If God is God surely Mohammed knew his nature when he appointed men to call by night, bidding the faithful rise to pray, and speaking as it were with Allah face to face, as standing on some tower in the night, they tell his attributes.

Chapter VI

NEXT morning rain was still coming down in
torrents and we awoke to find our tent, in spite
of all precautions, swimming in water. Nothing
to do, even without the " hospitality " of the Kaid, but to
cower over a charcoal brazier, and to send Swani to try
and buy provisions for our breakfast. After a little he returned bringing the Chamberlain, who informed us rather
tartly that the Kaid Si Taieb ben Si Ahmed el Hassan el
Kintafi* sold no provisions, but that we and our animals
would be cared for at his expense. In about half an hour
two negro slaves appeared, bringing couscousou and meat
cut up and stewed with pumpkins, and so we fell to with
an appetite improved by the past days spent on short commons. People arrived in bands and squatted down before
the tent, and when one band had seen enough, another
took its place and sat on doggedly for half an hour with
the rain dripping down their backs. Till about midday,
when the rain began to slacken, no one addressed us,
though a man, shod with sandals of oxhide with the hair
on, after the style of those used in the outer Hebrides till

* The Arabic name of the tribe is Kintafah. This in Shillah becomes Takinteft,
following the Shillah etymology (see Appendix).

a few years ago, and once worn by all Scotchmen, as the
name "rough-footed Scot" implies, looked at us, opened
his mouth, but then thought better of it and passed away.
Had he but spoken, it would have been in Shillah, but
still I wish I knew what that poor tribesman was about
to say.

After the midday prayer, a jet-black negro walked into
the tent, dressed all in white with a large silver earring
hanging from one ear and with three stripes tattooed or
scarified upon his cheek. He sat down in a friendly manner on my cloak, which I drew to me and he then sat quite
as contentedly on the wet mud. He spoke good Arabic,
and to my observation that it was cold, replied "Yes, but
not half so cold as London." Thinking it was a joke, I said,
"That is extremely likely, but have you been there?" and
I then learned he had been there twice, in the suite of a
Sheikh from the Sahara who had been in London about
some business of the Cape Juby Company. The Cape Juby
Company was one of those strange ventures which when
they fail, men say that their promoters were all mad, but
if they prosper make their projectors founders of empires,
and people think their brains effected that which was really
chance. There is no really sane head either on horseback*
or the Stock Exchange.

Almost in sight of Fuerteventura, the nearest in-shore
of the Canaries, between Wad Nun and the Cabo de Bo-

---

* *No hay hombre cuerdo á caballo* ("There is no sane man on horseback"),
says the proverb.

jador, not quite within the tropics, Cape Juby lies. From the Wad Nun, almost to St. Louis Senegal, nothing but sand, a lowish coast, no settled population, and but little water; and yet a Scotchman, one Mr. Mackenzie, who in the seventies had an idea to flood the Sahara and trade direct from London to Timbuctoo, induced some lords and gentlemen to found a company and put their factory on an island off the Cape.*

The island is a lone rock, disjoined from land by a little channel at high tide. Upon it, without much ceremony, the company erected buildings, built a wall, mounted the inevitable cannon to keep the peace, and tried to start a trade. Just opposite the island grow a few tamarisk bushes, in Arabic called Torfaia, and near these bushes is a brackish well. The tamarisks give the name, and the spot is called Torfaia up and down the coast, being remarkable as the only place where any vegetation lives. The population is nomad and composed of wandering Arab tribes, who in remote times came from the Yemen and still speak the dialect of the Koreish in all its purity. When the wind blows, the sand moves like a sea, so that the company found their walls, built for defence, were in about three months all sanded up, and an Arab on his camel could fire right down into the fort.

Upon this likely spot Mr. Mackenzie and his company

* A cousin of the celebrated Abd-el-Kader, one Haj Ali Bu Taleb, was employed by them to conciliate the natives. But though a man of ability and held a saint amongst the Arabs, he was unsuccessful.

pitched to establish trade with the interior of the Sahara, pending the time their ships could sail direct to Timbuctoo. If I remember rightly, the Empire (British) was to be increased, the wandering Arabs all to acknowledge Queen Victoria's reign, and peace and plenty, with a fair modicum of profit, be the order of the day.

Below Wad Nun the Arab tribes live practically as they lived in Arabia before Mohammed. Morocco is too far away to be much feared, the negro tribes to the south and east not strong enough to fight, the Tuaregs not disposed to attack men like the Arabs who can only give hard knocks; whilst in the Senegal the French, in spite of that consuming hunger for sand, as witness Southern Algeria, Tunis, and the desire they are said to have to make a sandy empire from Tripoli to Senegal, are not so foolish as to adventure, merely for glory, to attack the desert tribes. Thus any Sheikh, or Sherif, who has a little power becomes a king, after the style of Chedorlaomer in the Bible, and rules over other kings who do him homage, if he has power enough at his command. Some kinglet of this kind was duly found, and a treaty made with him. In fact, the man appeared in London, passed by the title of King of the Sahara, stayed at a good hotel, and no doubt did members of the company good service by his stay. But as mischance would have it, treason broke out at home whilst the Saharan King was sojourning amongst the infidel. On his return " patriots," that is, I fancy, other Shiekhs the company had overlooked and not " made right," refused to ratify the treaty, and the poor King was treated as a traitor who had sold his fellows to the Christian dogs.

Years passed away (sixteen I think), and managers went out who drew their salaries; clerks not a few contracted fever, and some died; Arabs attacked the place with varying success; the people of the fort went out to hunt in peace time, and in war sat drinking gin; but still no business was effected, and the whole place began to look like a small Goa, Pondicherry, or some Spanish island in the Eastern seas, where every other man has a fine gold-laced uniform, but has no food to eat. Then some one had a luminous idea, which was to sell the place to the Sultan of Morocco, who, wishing to extend his empire, bought it, and has today the pleasure of keeping a steamer running from the Canaries, week in and week out, to keep the garrison supplied with water, for the well beneath the tamarisks has given out.

My negro visitor was from Wad Nun, had been to London twice and did not like it, in fact, remembered nothing but the cold and the striking fact that the horses had short tails. Yet he was intelligent after his fashion, and greatly exercised as to the reason which impelled those Argonauts to sail to such a Colchis, and glad to tell how they bought nothing but a little wool, and wondered if they knew of gold mines, or what kept them stuck for sixteen years in such a place. The manager, Mackenzie, was a Scotchman, which he interpreted to be a title of respectability, and informed us that Mackenzie's name was known far in the Sahara, and that at times natives would ride up, who had never seen him, and greet him as Sidi Mackenzie, for they all had heard of his red beard, his title (Scotchman), and of the strong spirit kept in a barrel which none but he might drink. My negro tapped his head and told me it was full of news, but that, unluckily, his purse was empty, or he would long ago have left the place. I did not take the hint, and he retired wishing me patience, with a negro grin.

Patience was what I most had need of, except, perhaps, tobacco, at the time. So strolling out on to the Maidan, the rain having ceased at last, I sat down on a large stone under an olive tree to think about the situation and enjoy the view. In a magnificent amphitheatre of hills with snowy mountains towering overhead, just at the angle which the Wad el N'fiss makes after the first five miles of its course, about the middle of a valley almost ten miles in length, the Azib* of el Kintafi lies; around it fifteen or twenty acres of Indian corn, a grove of olive trees and pomegranate gardens, wild and uncared for as gardens always are throughout the East. The house itself with its mosque, the various court-yards, towers, kiosques, stables, fortified passages, and a long stretch of crenellated wall, covers almost as much ground as Kenilworth or Arundel. Built all of mud, and here and there painted light yellow, it yet looks solid, and in one angle rises the tower of the mosque covered with tiles like those of the Alhambra.

* Azib is a country house or farm.

The castle wall upon one side runs almost sheer down into the river which tosses on a stony bed, leaving a sort of sandy beach, on which grow oleanders, and across the stream the shoulder of Tisi Nemiri* almost reaches to the bank. Above the oleanders ran a mill stream in which the tiniest barble played or hung suspended in the slack water, aping the attitude of salmon lying suspended in the shallow water of a linn. Between the mill stream and the beach grew some Indian corn fenced in with suddra bushes, looking steely-grey in the bright sun, or as if frosted, when the moon turned every twig and thorn to silver in its cold light. Curious little round hills studded the valley in various directions, and on the west side rose Ouichidan to an apparent height of about fourteen thousand feet, or, say, six thousand feet above the bed of the N'fiss. Almost at the top of Ouichidan there is a spring held sacred by the Berbers, who have retained many Pagan customs and superstitions, although Mohammedans. It is said that from the top a view of both Morocco city and Tarudant is seen, but Allahu Alem, God he knows, for never has the foot of unbeliever trod the snow, nor has the pestilent surveyor, with his boiling tubes, his aneroid, theodolite, and all his trumpery, defiled the peak.

What is known, is that a pass leads into the Sus over a shoulder of the mountain, and a poor Jewish merchant whom I subsequently met, informed me that whilst crossing it he had prayed more fervently than he had done

* Tisi Nemiri, in Shillah, the Hill of the Stones.

since quite a child, and said devoutly, "May Jehovah keep me from all such cursed roads." Amongst the maize fields, which at my time of sojourning as guest of El Kintafi were all ripe, negroes and negresses were husking the heads of maize which had been reaped, and were all gathered up in heaps. Their flat and merry faces, red and yellow clothes, loud cries, and air of working as for amusement, brought back the Southern States, and as half the men answered to Quasi, and the women all appeared to be called Sultana, the illusion was complete. Most of the negroes had become Mohammedans, and, of course, the women had to follow suit; few of them spoke Arabic or even Shillah, and a sort of ganger, who spoke Soudanese, lay in the shade and made a show of overseeing their pretence of work. Upon the flat roof of the palace prisoners heavily chained tottered about and husked the Indian corn, each man resolved to do as little work as possible, and spend the greatest possible amount of time in walking to and fro. Poor devils, mostly tribesmen from a rebellious tribe which lived upon the head waters of the N'fiss, and which neither the Kaid, his father, or his grandfather, had ever been quite able to subdue. The rebellious tribe is a feature of the Eastern world. No Sultanate so small, no little caliphate lost in the hills, no territory of mountains or of plains that is without its rebels. Throughout Morocco one comes now and again upon a tribe in open warfare, if not with the Sultan at least against its Governor. They raise an army, fight in the hills, take prisoners

and cut their throats, behave, in fact, as Arabs have behaved since they first came into our Western life, and at
the last the Sultan or the Governor prevails, and a few
dozen heads of the chief tribesmen adorn the city walls,
making long smears of blood down the pale yellow wash,
and shrivelling by degrees into a hideous mass, like an
old fly-blown shoe. Yet on the roof of the Kaid's Kasbah*
the prisoners gave a scriptural note, and made one think
on Jeremiah,† the prisoner-prophet, who must have wandered up and down as they did, before Pashur smote him
and set him in the stocks, putting him daily to derision
and making everyone to mock at him. Not that the Berbers mocked at the prisoners on the castle roof, but, on
the contrary, sat with the blazing sun upon their shaven
pates, talking to them from the Maidan, gave them bits
of bread, and so behaved themselves that, chains and
famine, lice and sores discounted, the condition of the
prisoners was nearer to that of men at large than that of
the betracted, ticketed transgressors in a London gaol. At
times upon the roof the prisoners lay full length and held
their chains up in one hand to test their legs, and, lying
close to the parapet, chaffed the negroes as they came and
went carrying as little maize as possible in small baskets
holding about a peck. At night the Kaid, who had a not

---

* Kasbah = a castle, and from it is derived the Spanish word Alcazaba, so
frequent in Spanish place names.
† In a Bible in Pentonville Prison, at the end of the Book of Jeremiah, a prisoner
had written, " Cheer up, Jeremiah, old man." I used to be sorry for both prisoners
when I read that Bible in my cell.

unnatural wish to keep his prisoners safe, lowered them
one by one into a deep, dry well, a mule revolving slowly
round a rude kind of capstan, as with an esparto rope
hitched in a bowline below their arm-pits, one by one
they were lowered underneath the ground. When all were
down, four negroes placed a large flat stone over the well,
and the Muezzin called on the faithful all to praise God's
name. What the state of the well was down below is hard
to say, but in the morning when the stone was rolled aside
a stench as from a Tophet rose, and early on the fourth
day of our enforced "villegiatura," a starveling donkey
was driven past our tent with the body of a prisoner
(escaped from prison and from life) thrown over it, the
head and feet dangling upon the ground, and the donkey-
driver pricking his beast with a piece of sharpened cane
in an old, thoroughly-established sore over which the flies
buzzed, settled, flew off again and then alighted on the
eyelids of the corpse.

And so the second day went past in rather greater comfort than the first, the rain having ceased, and we became
aware that it was best to resign ourselves to what the
Kaid and Providence might have in store for us. About
the evening prayer I despatched Lutaif to try and get an
audience and find out why we were detained, but, after
waiting more than an hour at the castle gate, he returned
to say that the great man was invisible, but that his Chamberlain would come and see us after supper-time. The
Chamberlain not having come by sundown, and our ani-

mals not having eaten for four-and-twenty hours, I sent
Swani to "stand and cry at the gate," after the Eastern
fashion, with the result that soon the slave who had the
key of the corn arrived and struck a bargain with me, I
undertaking to tip him handsomely on my departure if
he fed our beasts. This he declared he would do, but kept
them half-starved, and totally refused to sell, either chopped
straw or hay, most probably having got orders from the
Kaid to feed our animals well, and having sold half the
corn, which he no doubt took every day out of the granary,
to passing travellers. However, unjust stewards are not
uncommon even in England, and he kept his key so bright
(it measured nearly eighteen inches long), and dressed so
charmingly in palest Eau de Nil, with a black cloak bound
round with pink, and lied so easily, and with such circumstance, that I forbore to lay his case before the Kaid,
reserving to myself the power of not rewarding his hypothetic services at my departure if he went too far in his
small villainies. Night brought the Chamberlain heading
a small procession of negroes carrying our supper, composed of various dishes of meat and couscousou, piled up on
earthenware dishes placed in a wooden case, shaped like
an old-time stable sieve and covered with a conical-shaped
top of straw through which were run strips of red and
black leather in a curious pattern forming a kind of check.
A dish of Moorish bread fresh from the oven and made
of brownish flour, well-garnished with particles of the
rough stones of the bread mill in which the corn was

ground, made up the banquet. We squatted on the floor, Swani went round and poured water from a tin dish upon our hands, and, after a pious Bismillah, we all dipped bits of bread in the red grease and oil of the highly seasoned dishes and began to eat, ladling the food into our mouths most painfully with the right hand, and lifting the huge dishes to our mouths to drink the soup. A loud *Bon jour* warned us the Persian was outside the tent, waiting to tell us his adventures and to impart the gossip of the place. Entering, he sat down with a sonorous " Peace be on you all," and, after one or two cups of tea and a few cigarettes, began to talk. The Kaid, he said, was puzzled about us, had thought at first of sending us in chains to the Sultan's camp, again had thought of letting us proceed, and yet again of sending us under a guard of his own men to the Consul at Mogador, but at the last had not imparted his design to our informant. He thought that probably a mes- senger had been despatched to the Sultan in his camp, and if that was the case we had better make our minds up to at least a fortnight in the place, and afterwards, if the Sultan should send for us, a fortnight at the camp, for things in palaces go slowly;* but, said the Persian piously, " may God not open the door of the Sultan to you, or to any other man." By this I took it that should the Sultan send for us we should pass an unpleasant quarter of an hour, though as there were several Europeans in the camp, notably Kaid M'Lean, the instructor-general to the army,

* Coses de palacio van despacio.

and a French doctor called Linares, we might be able to get things arranged upon the peace-with-honour plan. This matter duly gone into with proper Oriental deliberation, the Persian entered on his tale. Born sometimes in Shiraz, at others in Tabriz (according to his fancy), it appeared that at an early age he had left his country as an Ashik,* that is, a wandering singer, and gone to Turkey. There he had acquired the Turkish language as he averred, but subsequent intercourse with Arabs had mixed the tongues; in the same fashion a Spaniard, Italian, or Portuguese gets mixed, on learning either of the kindred languages. Such as it was, his jargon suited me, and I spoke Arabic more fluently to him than I ever spoke that tongue before, or think to speak it until I meet another Ashik similarly graced. We spoke without a verb, without a particle, like idiots or children, largely in adverbs and in adjectives, and without shame on either side, each thinking that the other was but little versed in Arabic, and that he condescended to adopt a jargon to help his weaker friend. The Persian's faithful fiddle was out of order, owing, he said, to his falling off a donkey whilst on the "roads of hell" which led up to Kintafi, but though without it, he sang for hours, songs which he said treated of love in Persian, and which I took on trust as not containing anything subversive of morality, or fit to raise a blush upon the cheek of those used to our Western ways. Hafiz he said

* "Ashik" literally means "lover" in Persian, and has by degrees come to mean a minstrel, because in the climate of Persia lovers are assumed to sing to lutes and other instruments, which the climate renders unseemly in England.

he knew by heart, and much of Jami, and those Rubaiyat which every weekly newspaper has its columns full of he knew, but found them too materialistic for his taste.

In outward visible appearance he was a perfect Kurd, squat and broad-shouldered, his beard so thick that had you dropped a pin in it it would not have touched the skin, eyes black and piercing, face like a walnut, and a long love-lock hanging from his shaven head down on his shoulder; in one ear he wore a silver earring, and he was gifted with a voice perhaps the strongest that I ever heard in man. Both a musician and a philosopher, but yet a moralist, and a fanatic preacher of the graces of Mohammed, but still a sceptic at the heart, and above all a traveller, " for by travelling a man, although his purse grows light, still lays up treasures on which to live when he is old. What is a Sultan, Kaid, Pasha, or Governor, compared to him who is the Sultan of the world, when where night catches him he rests and looks upon the stars?"

So, starting from Shiraz (occasionally Tabriz), he had run over almost all Europe, Turkey, and a portion of the Northern States of Africa. Had been in Serbistan, Atenas, and in Draboulis (Tripoli), Massr-el-Kahira (Cairo), Stamboul, and Buda Pesth, and of all lands in Europe thought most of Magyarstan; " for there the women have eyes like almonds, though they drink too much beer, the men are tall and fierce, handsome (husnar besaf), the horses large and elegant, they run as lightly as gazelles on stony

ground; and had the people but the blessing of the true
faith, he never would have left their land." Their city,
Buda Pesth, was larger than Stamboul, more fine than
Paris, Vienna, or than Shiraz, and in the middle ran a
river on which went steamboats, in which he used to travel
and pay his passage with a tune, for the Magyars, he said,
loved music better than they loved their God. In this
famed city he had known one Bamborah, who by interior
evidence and after cogitation I found to be Professor Vam-
bery. Large-hearted was this Bamborah, and speaking Persian, a Christian dervish, knowing all the East, having read
all books, explored all countries, mastered all sciences and
learning; the friend of kings, for had not the Sultan Abdul
Hamid (whom may God preserve) sent him a ring of
" diamont " worth a thousand pounds, and Bamborah had
shown it whilst they sat discoursing in his hospitable house.

In fact of all the men, Christian or Moslem, he had ever
met, this Bamborah appeared to him the fittest to stand
before a king. But for himself even in Buda Pesth, the
travelling fever had impelled him to embark aboard a
" chimin de fer," and go to Vienna, travelling all through
the night and reaching Vienna about the feyzir (day
break), and straying up and down the streets until at last
he met a Turk who sold red slippers, and lodged with
him; but after several days spent in the place, which gave
him neither pleasure nor material gain, went on to Baris
to find upon the journey that the customs of the East and
West did not agree. The Christians know no God. With

them it is all money (Kulshi flus), with them no stranger, no wayfaring man, for in that train to Baris he asked a woman for some water to wash his hands so as to address Allah after the fashion laid down in the holy books; she brought it, and after washing, and his prayers all duly said, the passengers, as he informed me, crowding about in an unseemly way to see him pray, he smiled and thanked the woman, and taking out a cigarette tendered it to her with his thanks. But she, born of a dog, knowing no God and dead to shame—for is it not set down " to strangers and to wayfarers be kind"?—laughed an ill laugh and asked for a half-a-franc: franc, franc, and always franc, that is the Christian's God.

Thus talking of the Alps and Alpujarras the time wore on, and after saying again most earnestly "May God not open the door of our Lord the Sultan to you," he took his leave, and as he went it seemed our only friend had gone.

Next morning found the situation still unchanged, and we began to look about us and found out that we had several companions in adversity. Camped in a tent about a hundred yards away were three Kaids (*Biblicé* kings), from the province of the Sus, who had been waiting more than a month for an audience of our captor, he having summoned them to wait upon him to confer, as the gipsies say, about "the affairs of Egypt." One of the "kings" turned out to be a "saint" of some repute, a tall fine man of Arab type and race, dressed all in spotless white, and reading always in a little copy of some holy book under an olive tree, showing no trace of trouble at his long wait, although he must have passed through much annoyance and incurred considerable expense, as almost all his animals had died through lack of food and the change of climate from the warm lands of Sus to the cold winds of the interior Atlas range.

Under the olive tree I sat and talked to him, chiefly through the medium of Lutaif, and asked him much of Tarudant, from whence his house was situated but a long day's ride. It appeared that in the main the account of Gerhard Rohlfs,* the Hamburg Jew, who visited it some thirty years ago, is applicable to the city of today. The Sherif spoke of the high walls mentioned by Rohlfs and Oskar Lenz;† of the five gates called Bab-el-Kasbah, Bab-el-Jamis, Bab-Ouled ben Noumas, Bab Targount, and Bab Egorgan; of the high Kasbah, occupying, as Rohlfs says, a space of 50,000 square metres, and cut off from the town by a high crenellated wall. He dwelt upon the cheapness of provisions, said that six eggs were bought for a little copper coin called a "musonah" (known to the Spaniards as "blanquillo") and worth perhaps a farthing. He said a pound of meat cost two or three "musonahs," spoke of the trade in brass and copper vessels, and gave us to understand, of all the towns within the empire of his Sherifian Majesty, that Tarudant was cheapest and pleasantest to live in, and inferred it was because the people had no

* Gerhard Rohlfs's "Adventures in Morocco." London, 1874.
† Oskar Lenz, "Timbouctou." Paris, 1886.

dealings with the infidel. For, said he, " the infidel are
Oulad el Haram " (sons of the illegitimate), ever a-stirring,
never contented with their lot, afraid to be alone,
seeing no beauty in the sun, not caring for the sound of
running water, and even looking at a fine horse but for
his worth; men bound to a wife, the slave of all the things
they make; then, recollecting that I too was one of the
dog descended, he gravely drew his hand across his beard,
and said, "but no doubt Allah made them cunning and
wise for some great purpose of his own."

No doubt, in every town throughout the East, the presence of even a small quantity of Europeans forces prices
up, upsets the national life, unsettles men, and after having
done so, gives them no equivalent for the mischief that
it makes.

The mosques were three in number, one in the Kasbah,
two in the town, of which the principal was El Djama-
el-Kebir, and the most sacred Sidi-o-Sidi, the Saint of Saints,
by which name people of a serious turn of mind call the
whole town in conversation, as who should say, speaking
of London, the city of St. Paul.

I questioned him about the sugar-cane plantations, of
which both Luis de Marmol* and Diego de Torres† speak

---

*Luis de Marmol y Carbajal was a prisoner in Fez, and wrote a curious book,
called, "Descripcion General de Africa." He was born in Granada in 1520,
accompanied Charles V. in his expedition to Oran, served in Africa twenty years,
was made prisoner, and remained eight years in Fez. On his return he wrote his
book.

† Diego de Torres (a Valencian), " Relacion del Origen, y Suceso de los
Xaribes, y del Estado de los Reinos de Marruecos, Fez," etc. Seville, 1584.

in their curious books written in the middle of the sixteenth century, but he had never heard of sugar-cane near
Tarudant. This forms another proof of the decadence into
which all the land has fallen, for the climate and soil of
Tarudant should be at least as favourable for sugar-cane
as is the strip of territory in Spain which runs along the
coast from Malaga into the province of Almeria, where
the Moors first introduced the canes which grow there still
today.

It appeared according to the Sherif that the town contained some 1,400 houses, and a population which he estimated at 10,000* mostly in easy circumstances, but very
ill-disposed towards all foreigners. Much information he
imparted as to the mineral riches of the province; but
without descending to particulars, with the exception of
sulphur, which he said was found close to the town. Romans, of course, had left their castles near the place, in
the same way the Moors have done in parts of Spain to
which they never penetrated; but who those Romans were
remains for some more well-graced, or fortunately starred
traveller than I to tell of, commentate upon, and weave
a theory to content his public and himself. Caravans, it
seems, go straight to Timbuctoo taking European goods,
and bring back slaves and gold-dust, with ostrich feathers
and the other desert commodities, as in the time of Mungo

---

* Joachim Gatell, " Description du Sous," " Bulletin de la Sociéte Geographique
Paris," sixième serie, 1871, pp. 86-89, puts it at 8,300 or thereabouts, and the
houses at 1,300.

Park. At least they did; but I suppose the French in their consuming zeal for freedom may have stopped slaves from being bought in these degenerate days. I learned the chief fondak or caravansary was kept by one Muley Mustapha el Hamisi; but most unluckily an unkind fate deprived me of the opportunity of entering his hospitable walls. The city seemed to resemble, from the account of the Sherif, Morocco city, that is, to occupy a relatively enormous space, as almost every other house had a large garden, and several of the larger houses gardens of many acres in extent; so that, as the Sherif explained, " the town looks like a silver cup* dropped in a tuft of grass." This, as I did not see it, I take on trust, believing perhaps that Moses had died happier had he not had the view from Pisgah's summit over the plains of Canaan.†

And so the Governors waited on patiently, their followers almost starving, and I expect themselves not too much fed, for the Kaid's servants ate or sold all the provisions which the Chamberlain issued each day to feed the various " guests." How the poor devils in the prison underground fared I do not know, but now and then a bucket with bread

* The Spaniards call Cadiz "Una taza de plata," a silver cup.

† It is, however, a reproach to our travellers that this town, accessible to travellers of all nations in the sixteenth century, is less known than Mecca today, and has never been visited by an Englishman this century. Mr. Walter Harris, the well-known traveller in Morocco, made a much more difficult and dangerous journey to Tafilet, and if he essays Tarudant, I wish him the success which in my case was withheld by Allah. The inhabitants are, without doubt, the most fanatical in Morocco; but I am certain that, had I had more time for preparation and an adequate knowledge of Arabic, I could have both reached the place and come safely away. As it was, I had no one to consult with, little time at my disposal, and I knew little Arabic, and that little badly.

and couscousou was let down to them, and I believe they flew on it like half-starved jackals on a dead donkey outside a Moorish town. Although a semi-prisoner, the Sherif from Sus was still a holy man, and therefore I sent Mohammed el Hosein to take refuge with him till all talk of fetters, dungeons, stick and shaving beards was past. This taking refuge (to " zoug," as it is called) is common in Morocco. At times upon a journey, some man will rush and seize your stirrup, and will not let go till you promise him protection, which, when you do (for it is not easy to refuse), you make yourself an Old Man of the Sea, an Incubus, who has proprietary rights in you, and perhaps follows you to your journey's end.

The patience of the three Sheikhs gave me an example, and I endeavoured to take our enforced detention as quietly as they did, though without success. It was truly wonderful to see them sitting all day long, half-starved, outside their tent, taking the sun and praying regularly, yet without ostentation, telling their beads, listening to the Sherif read in a low voice from his little sacred book, and praising Allah (I have no doubt) consumedly.

And so the day went past without an incident, except that towards evening, as they drove the cattle home, two tame moufflons came with them, as goats do in a field at home. These moufflons (Oudad) are rather larger than a large goat, almost the colour of a Scotch red deer; a tuft of hair about a span in length grows on their withers and on their dewlap is another of the same size; the eyes are large and full, and very wide, and their chief feature is an enormous pair of curving horns. A wandering Sherif (in this case a kind of fakir) from Taseruelt told me that "they often throw themselves off a cliff a hundred feet in height, and fall upon their horns, and then jump up at once, and run off faster than gazelles run when the hunter shouts to his horse." Just before evening we met the Kaid's secretary, a quiet, handsome, literary man, who came and sat with us, and talked long about books, and grew quite friendly with Lutaif on his producing the poems of el Faredi.* They read aloud alternately in a sort of rhythmic sing-song way, pleasing to listen to, and which is taught in Arab schools. Though I did not understand more than a word in five, the language is so fine, I enjoyed it more than all the matchless eloquence of a debate in Parliament.

To read in Arabic is a set art; to read and understand a different branch of scholarship; but these two Talebs both read and understood, and after an hour's intonation, strophe about, they marvelled at one another's learning, and like two doughty chieftains in an Homeric fight, stopped often to compliment and flatter one another. " By Allah, it seems impossible a Christian can read and understand el Faredi." " Strange that a Taleb of the Atlas should know the literal† language as if he were an Eastern "‡ and the like. The " Taleb of the Atlas " explained he had been the pilgrimage,

---

* El Faredi, born in Cairo, 1181-1235 (Christian era). His only work which has survived is a collection of poems in praise of Allah, known as "El Divan-el-Faredi." Little is known of him but that he was a fakir.

† "Literal " as opposed to the spoken or "vulgar " Arabic.

‡ " Eastern " carries with it something of holiness, as born near the holy places.

and lived two years in Mecca, and whilst there (although a Berber) had studied deeply and perfected himself in the knowledge of the East as far as possible. Lutaif explained that though a Christian he was of Arab race, and that he worshipped God, as " Allah," in the same way as did Mohammedans. " Then," said the Taleb, half laughing, " either you are a Moslem in disguise, or else a Taleb who has become a Christian." Seeing the conversation was becoming rather strained, I interrupted and broke it up; but when the Taleb left, Lutaif borrowed my knife and managed to haggle off his beard, though not without abrasion of the cuticle, and though without it he looked less like an Arab, still with his moustache, which he refused to sacrifice, he looked so like a Turk that, as regards appearance, little was gained by all his sufferings.

The next day found us with the same postal address, still without having seen the Kaid, and without a definite idea of his intentions. Most of the people seemed to be certain that a messenger had gone to the Sultan for instructions on our case, but both the Chamberlain, the Taleb, and the Captain of the Guard denied with circumstance, and perjured themselves as cheerfully, and with as much delight in perjury for the mere sake of perjury, as any minister answering a question from the front bench of the grandmother of all parliaments. We passed the time reading el Faredi and an Arabic version of the Psalms; writing and smoking, walking up and down the Maidan, sitting underneath the trees, and watching the Kaid's horses and mules being driven to the river to bathe and drink. Although a Berber and a mountaineer, the Kaid was fond of horses, and had a stud of about eighty horses and as many mules. Negroes led down the horses all " lither-fat," for our lord the Kaid had " long lain in," and there had been no riding in the glen for the past month. Blacks, bays, and chestnuts, with a white or two, and a light cream colour of the kind called by the Spaniards " Huevo de pato," that is duck's egg. All rather " chunky," as the Texans say, some running up to about sixteen hands, mostly all with long tails sweeping upon the ground and manes which fell quite to the point of the shoulder in the older horses. Their tails all set on low (a mark of the Barb breed), their eyes large and prominent, heads rather large, ears long, thin and intelligent, always in motion, backs rather short, round in the barrel, and well ribbed up, straight in the pastern, and feet rather small and high, the consequence of being bred on stony ground.

I learned the black* (el Dum) is best for show, but bad in temper, especially if he has no white hairs about him. Ride him not to war, for when the sun shines hot, and water is hard to find, he cannot suffer, and leaves his rider in the power of his enemy. Still the black without moon or stars (white hairs) is a horse for kings, but he fears rocky ground. The chestnut† when he flies beneath the sun, it is

---

* The Spaniards also say a jet-black horse (zaino) is bad tempered.
† " Remarks on Horsemanship," by the Emir Abd-el-Kader.

the wind. It was a favourite colour with the Prophet, and
therefore to be desired of all good Moslems and good
horsemen.

The roan is a pool of blood, his rider will be overtaken
but will never overtake. The light chestnut (Zfar el Jehudi,
the Jews' yellow) is not for men to ride, he brings ill luck.
No wise man would ride a horse with a white spot in front
of the saddle, for such a horse is as fatal as the most violent
poison. In the same way no prudent man would buy a horse
with a white face and stockings, for he carries his own
shroud with him.

The white is a colour for princes, but not for war. When
you advance afar off your enemy makes ready for you.

The bay is the pearl of colours, for the bay is hardiest and
most sober of all horses. Says the Emir Abd-el-Kader, " If
they tell you a horse jumped down a precipice without
injury, ask if he was a bay, and if they answer yes, believe
them." Lastly, the Emir says with reason, " Speak to your
horses as a man speaks to his child, and they will correct
their faults which have incurred your anger, for they understand the mouth of man."

Armed with these maxims and on the look-out for others,
I was not dull as long as the horses were in sight. Sometimes a boy would ride them in to swim in the swift current,
snorting and plunging till they lost their feet,   then
their heads appeared out of the water, their backs almost
awash, the boys clinging to them like monkeys as they
struck out for the bank, raising a wave like small torpedo

boats. At other times two would break loose and fight, screaming and standing up, or rushing in, seize one another by the necks like bulldogs, when their respective negroes dodged outside, like forward players in a football scrimmage waiting for the ball, trying to catch their ropes but afraid to venture in between them. Generally the fight was ended by an Arab or a Berber rushing up armed with a thick stick and a handful of round stones, with which he beat and pelted them till they let go their hold. Others again would break loose all alone, and career about the sandy beach, head and tail up, or gallop through the corn, their attendant running after them in agony till they were captured. The most sedate walked delicately as they were Agags down to the water, plunged their muzzles deep into the stream and drank as if they wished to drain the river dry, looked up and drank again three times, then turned, and after executing several perfunctory bounds, lay down and rolled in the wet sand and quietly walked home beside their negro, not deigning even to look at any other horse, then disappeared under the horse-shoe gateway to the inner courtyard where they lived. The mules were not so interesting, though valuable, fetching more money than the best horse, and if accomplished " pacers," often bringing two or three hundred dollars, whereas a horse rarely exceeds a hundred and fifty at the most; but still they have an air as of a donkey, which makes them quite uninteresting, for they are lacking in the donkey's inner grace.

In districts like the Atlas mules are more serviceable than

any horse, and on the mountain roads will perform almost a third longer journey in a day. Where the horse beats them is on the plain, for no mule can live beside a horse at the horse's pace, though on a rough road the mule's pace is much the faster of the two. Sometimes five or six mules would break loose and follow one another in a string, jumping the thorny fences heavily, their ears flapping about ridiculously, and their thin tails stuck high up in the air. They never swam, and, on the whole, had I been limited to mules exclusively, I could not have passed my time so well. One horse especially interested me, a large creamy-white animal with an immense and curly mane; he always came alone, led by two negroes, and had an open wound upon his head and two upon his chest. I learned he was the Kaid's special favourite, and that about a month ago, during an expedition to the Sus to aid the Sultan against a refractory tribe, the Kaid had received a bullet in the leg, and the horse had got his wounds at the same time. None of the horses that I saw would be of any value, except to an artist, in the European market; but for the country where they were bred they were most serviceable, hardy, and indefatigable, sober beyond belief, eating their corn but once a day, drinking but once, and up to any weight, and if not quite so fast as might be wished for, still a glory to the eye.

The horses gone, the entertainment of the day was over, and I got quite accustomed to expect them at a certain hour, and to be quite annoyed if they were late.

Thus did one day tell and certify another, leaving us quite

cut off from all the world, as far removed from European influence as we had been in the centre of the Sahara; well treated, but uncertain of how long we should be kept in honourable captivity, growing more anxious exery moment, and yet with something comic in the situation; nothing to do but make the best of it, eat, drink, and sleep, and stroll about, talk with the natives, sit in our tent, and read el Faredi, giving ourselves up with the best grace we could, to watching and to prayer.

Chapter VII

THE 24th still found us, so to speak, in Poste Restante at Kintafi, the Kaid invisible, tobacco running low, food not too regular, and our animals becoming thinner every day. Still the example of the prisoners, the Sheikhs from Sus, and a tent full of miserable tribesmen, all almost without food, and glad to eat our scraps, kept us for shame's sake patient. So we talked much to everyone, especially with the negro who had been in London, and found he was a man of much and varied travel, some experience, no little observation, and ready to talk all day on all that he had seen. London had not impressed him, or else impressed him to such purpose that he was dumb about it; but of the Fetish worshippers below the Senegal he could tell much. In speaking of them, though a negro of the blackest dye, he treated them as savages, being a Mohammedan, and laughed at their religion, although the most foolish portions of it seemed to appeal to his imagination, in the same way that negroes in America (all Christians, of course), are seldom pleased with moderate Christianity, but usually are Ranters, Bush Baptists, or members of some saltatory sect, which gives them opportunity to enter more fully into communion with the spirit of the thing than if they sat and listened to a prayer, slept at a sermon, or dropped their money in the plate, merely conforming in a perfunctory way, as their less animistic lower-toned brethren in the Lord seem quite content to do.

Fetish, our friend explained, is good, and works great deeds; sometimes a man will die, and then the Fetish man appears and cuts off a cow's tail, fastens it on the dead man's forehead, who at once gets up and walks to where he wishes to be buried, and dies again. Why he does this the speaker did not know, or why, before he dies, the man does not explain in the usual way to his friends where he wants them to lay him after death. Nor did he know what good the fetish man receives by his operation, or if the tail should be taken from a dead or living cow; but he was certain that he had seen a miracle, and told us plainly that he never was so certain of anything before. Fetish to him seemed rather an incident than a religion, for he went on to say the heathen negroes have no God, just like the Christians; and then turned grey, which, I think, is the negro way to blush, and said, he did not point his observation at ourselves, for he had heard in London that we worshipped several Gods, and that the Christians he referred to were the people in the Canaries who, he was positive, worshipped a goddess, for he had seen them do so in a Mosque. Again he said his head was full of news, but his purse still continued light, and so he drew it out and showed it to us, and it was empty certainly; but beautifully made, of a most curious pattern and workmanship, and cunningly contrived of pieces of thin

leather which all fitted into one another and drew out, after
the fashion of those painted boxes which used to come from
India, which in one's childhood when one opened them, one
always found another underneath. It appears he has been
often in the Canaries, and knows the islands, as Lanzarote
which he denominates " Charuta," Fuerteventura which becomes " Fortinvantora," and Grand Canary, where he saw
one Christian kill another with a knife.

In form he was almost perfect as a type of race; blacker
than shale, with yellowish teeth like fangs, nostrils as wide
as a small donkey's, huge ears like a young elephant's, and
bloodshot eyes, thin, spindle legs, and all his body covered
with old scars; for he had been in many wars, " shoots," as
he said; " plum centre " rides well, and to crown all had
feet about the size and shape of a cigar box, stuck at right
angles to his legs, so when he walked he looked like a
flamingo, or a heron in a swamp. Knowing by actual
handling the exiguity of his purse, I approached the negro
to try if he would carry letters for me to the outside world.
But after having bargained for five dollars, a sudden panic
took him, and he refused, and ever after during our stay
avoided us, although I did not hear that he informed the
Kaid or any of his men.

All the day long a constant string of people kept arriving
at the castle gate. Little brown sturdy-legged Berber tribesmen, armed with long quarter-staffs, dressed in their dark
" achnifs," barefooted and bareheaded, save for a string of
camel's hair bound round the forehead; small-eyed, and

strangely autochthonous in type, as if they and the stones
upon the hills had sprung into existence long before history.
Small caravans of donkeys carrying Indian corn, with fruit,
with almonds, and with meat entered the gate full laden,
and came out empty. Negresses walked down the hill tracks,
bending beneath the weight of immense loads of brush-
wood, or of grass, or the green leaves of Indian corn or
sorgum,* to feed the horses, and to heat the ovens of the
Kaid.

I found that, though the Kaid oppressed and plundered
all the district, his oppression was in a measure balanced by
his charity, for he fed all the poor people of the valley, and
dispensed his hospitality to all and sundry who passed his
gates. So that, take it for all in all, his tyranny was only
different in degree from that of the manufacturer in the
manufacturing towns of England, who lives upon the toil
of several thousand workmen, discharges no one useful
function to the State, his works being run by paid officials,
and he himself doing nothing but sign his letters, whilst he
uses the money wrung from his workmen to engage in
foreign speculations, to swindle the inhabitants of distant
countries; and for all charity subscribes to missions to convert the Jews, or to send meddling praters to insult good
Catholics in Spain.

The Kaid, Si Taib el Kintafi, lives like a veritable prince,
is almost independent of the Sultan, and reminds me of a

---

* Sorgum is the Sesame of the Arabian Nights. In Spanish it is known as
Ajonjoli, a corruption of a word of similar sound in Arabic.

Mohammedan Emir in Spain after the Caliphate of Cordoba broke up. These Emirs were called by the Spaniards Reyezuelos de Taifas,* that is, Kings of a Section, or Hedge Kings; they had their courts in Jaen, in Malaga, in Almeria,† and in the South of Portugal. But in especial it recalled Jaen, being situated amongst the mountains, as that kingdom was with towering Atalayas (watch-towers) on every hill, and the Kaid's palace in the centre of the land. Guards were on every road patrolling within sight of one another. About five hundred negro slaves were scattered in the various castles. The Kaid kept all his money in iron boxes underground, and all his wives were guarded by gentlemen of the third sex,‡ so that the parallel between Kintafi and Jaen was almost perfect. For enemies he had the Sultan and every other Kaid of equal strength, but the impenetrable nature of his territory made him almost impregnable, and the armed soldiery, who lounged about in numbers round the castle and at every fort, made him secure as long as he could pay their services. The difference between a Spanish-Moorish mountain State and the Taifa of Kintafi was (though not apparent in externals, in policy, and in ideals of government), in other matters, very striking, for in the smallest Moorish Courts of Spain arts, sciences, literature, and general culture all flourished, and were encouraged by the kings.

---

* Taifa means a Band or Company. The word is more used in the East than in Morocco, in the Arabic of today.

† Almeria was one of the few cities in Spain founded by the Moors. Its name in Arabic is El Merayeh, the Looking-Glass.

‡ Someone, I think, has called journalists " gentlemen of the third sex," but these guardians were connected with no newspaper.

Poets, musicians, doctors, and men of science flocked to the capitals; both Almeria and Malaga were centres of culture, and the works of many writers who flourished in those cities still survive. But in Kintafi all was in decadence; for music, only a little thrumming on the Gimbri,* beating of tom-toms, an occasional song sung to the lute, and the shrill music of the Moorish pipe known as the Ghaita, and which survives in parts of Spain under the name of Dulzaina. In both countries it is always accompanied by a small handdrum.

Literature, in the modern Moorish State, is confined but to the Koran, the sacred books, an occasional Arab poet, and the study of the one book known to be written in the Berber tongue, and called El Maziri, which deals exclusively with the ceremonial of the faith. Strange that a people like the Moors, still brave, so fine in type, ardent in faith, sober in habit, and apparently (if history lie not more than usually) so like to what they were externally, when they shook Europe, should have fallen into such absolute decay. Literature, art, science, everything is forgotten; architecture but a base copy of their older styles, though still not so degraded as our own; bad government cries aloud in every town, on every road corruption runs rampant, and still the people, as mere men, are infinitely superior to any race in Europe I have seen; patient, and bearing hunger and oppression with a patience to put a saint to shame, and yet incapable of

---

* The Gimbri is a diminutive mandoline, the front of which is of parchment. It has only three strings, and yet its sound is less unpleasant than that of more pretentious instruments I have heard.

striking a blow to free themselves from the base tyranny which all must see and feel.

No man in all Morocco to rise and say, as the Sheikhs said to Omar when he asked if any one had a complaint against him: " By Allah, if we had had cause of complaint against thee, we had redressed it with our swords." And still, in spite of decadence, I never met a Moor who knew anything of our European life and institutions, except, of course, the servant class and a few men in place, and money-lenders, who wished to change their decadence for our prosperity upon the terms of bowing to a foreign race. They know that to the individual man, with all its faults, their life is happier than ours—somehow they feel instinctively that when we talk of freedom, liberty, and of good government, these mean freedom, good government and liberty for ourselves, and that they, by submitting, would become slaves, receiving nothing in return. Still they accept our gin, our teas, our powder, and in general, all the good things which Europe sends them, with civility; but decline to see that we, for all our talk, are any better than themselves. Baulked of our negro, I cast about to find some other stranger willing to earn a dishonest penny by carrying letters for the prisoners of the Kaid. After much talking and long negotiation over-night, a man made up his mind to venture to Morocco city, a journey of about three days, for the magnificent guerdon of four dollars, one dollar to be paid at once, and the other three upon arrival, either at Mr. Nairn's, the missionary's, house, or that of Si Bu Bekr, a well-known merchant, and for a long time British Agent in the town, or at the house of the Sherif of Tamasluoght, a protected British subject, whose Zowia is about ten miles outside the town.

The letters duly written, with much care and secrecy, some to our friends, others to consuls, one to the British Minister in Tangier, and two* to newspapers, were wrapped up carefully in a piece of strong French packing paper, tied round with a palmetto cord, and given to the man, who crawled up to the tent door at dead of night, whilst we sat waiting for him like conspirators. At the last moment, he demanded "one more dollar in the hand," dwelling upon the risks he ran, and which, indeed, were great. Not being prepared to bargain at such a time, I put the dollar in his hand, and bade him go with God, telling Mohammed el Hosein to explain in Berber all that he should do. They seemed to talk for hours, though I suppose it was not really long, and after having received instructions, minute enough to confuse a lawyer, he slipped away, crawling along the ground (his single long white garment looking like a shroud), until he reached a bank beside the mill stream, over which he dropped.

We sat late, talking over our tea, smoked the last of our tobacco, and calculated how long it would take to get an answer, if all went well, and if our messenger escaped the perils of the road. Mohammed el Hosein thought in five days, if in the interim it did not rain, and the rivers rise, we might have word, " for," he said, " when the man arrives, either

* These two are to be found in the notes, in extenso.

the missionary or Bu Bekr, or both, will send off messengers both to the Consul, to the Sultan, and another to the Kaid to tell him where we are. The messengers," he said, " having their own spurs and other people's horses, will not spare either, so that upon the evening of the fifth day we shall hear, Inshallah, if all goes well." All the night long we slept but little, fearing to hear shots and see our guide brought back escorted by a guard, but all passed quietly, and we congratulated ourselves on his escape, as we knew well the first ten miles of the road would be most difficult for him to pass.

Next morning we kept wondering, like children, how far our messenger had got upon the road, and trying to encourage one another by talking of the tremendous distance an Arab " rekass " could go when pushed to exert himself by fear or gain. The stories grew as the morning passed, until at last from eighty to a hundred miles seemed quite an ordinary trot. Privately, I thought the thick-legged mountaineer did not look like a record breaker, but I kept my opinion to myself, and let my followers yarn, seeing that it relieved their hearts. By this time, Mohammed el Hosein had ventured back from his refuge with the Sherif from Sus, and was quite sure that the Kaid had sent a messenger off to the Sultan, for, on the morning after we were captured, a " rekass " was seen to strike across the hills. The Chamberlain,* Si Mohammed, honoured us with a visit, wrapped up

* I render the Arabic word " Hajib " by Chamberlain, though I am not quite sure whether Chamberlain does not better correspond to the Moorish dignity of Kaid-el-Mesouar.

in fleecy, voluminous white clothes, new yellow shoes, beard
dyed with henna, finger nails turned orange with the same
substance, and followed by two stalwart, well-armed mountaineers, who strode behind him as he waddled to our tent.
After due compliments and salutations, he officially informed
us that the Kaid was wounded, and could not see us for a
day or two. Though we knew this before, to show good
manners, we pretended great astonishment, asked who could
have dared to wound so good a prince, where it had happened, and as to seeing us, begged he would use his pleasure
in the matter, though, of course, we counted every minute,
till we should see his face. The Chamberlain, as excellent a
diplomatist as could be wished for, said he was glad to see us
in such spirits, for he had thought by my expression on the
first day, that I was irritated with his lord, the Kaid, who
had detained us (as he now said) merely to save our lives,
as, had we crossed into the Sus, the " Illegitimate Ones "
(Oulad el Haram) would have slain us, not for Christians,
"for," said the Chamberlain, looking at me with a slight
smile, " you wear our clothes as if you had been born
amongst the Arabs," but as strangers, for God had made
their hearts like stone to all they did not know. The conversation languished till he startled me by asking if I was
a Moskou, which at first I did not comprehend, and thought
he meant to ask me of what sect I was. Just as I had instructed Lutaif to inform him I was a member of the U.P.*

* U.P. in N.B. stands for United Presbyterian. Stevenson refers to this sect, in
one of his ballads in the Scottish dialect. This sect is little known amongst well-
informed people, and its tenets have been greatly misunderstood.

Church, and was as orthodox* a Christian as he was a Mohammedan, it dawned upon me that he meant to ask me if I was a Russian, and might have said I was one, had I not feared to bring about a diplomatic question of some sort. Our guns against the tent-pole next struck his eye. Mine was a double-barrelled gun lent by the British Consul in Mogador. Lutaif's, a single-barrelled, Spanish gun, made in Barcelona, and so light upon the trigger, that on the only occasion he tried to fire with it, it started before he was expecting it, and the whole charge passed close to the head of an old negro on a donkey passing along the road. The negro did not turn a hair, but was just starting to abuse Lutaif, when, in my capacity of a Sherif, I rode up and cursed him for a dog for getting between the gun and the partridge Lutaif was firing at. This made things right, and the poor man rode on with many apologies for having frightened us. The Chamberlain examined the two weapons, and asked leave to show them to the Kaid, and, before we had time to answer, motioned to one of his attendants to take them off. I sat as solid as a rock, and said it was not usual to deprive a guest of his gun without consent, where- upon the Chamberlain poured out a torrent of excuses, and said the Kaid had now determined not to hurt a hair of any of our heads, and only wanted to see the guns from

* In Morocco the people belong to the sect of Malekiayahs, one of the four sects into which orthodox Mohammedans are divided, the other three being the Hannafiyahs, Shafiyahs and Hambaliyahs. Thus, had I declared myself a member of the U.P. church, I fear I should still have been less orthodox than the Chamberlain.

curiosity. Seeing the turn events were taking, I tried to get the Chamberlain to say if a " rekass " had been sent off to the Sultan or not, and if it was not possible for him to use his influence (for a consideration) to enable us to go on to the Sus.

Tipping through the East is understood to the full as well as it is understood in England, and his eyes glistened at the word " consideration," but with an effort.he replied, as far as he knew no messenger had gone, and, as to going to the Sus, had he not told us of the Sons of Belial who in that land cared for men's lives as little as an ordinary man cared about killing lice? After some tea and a surreptitious cigarette which I found in the corner of my saddle-bag, and which to make him smoke we had to close the tent, so that no one should see him " drink the shameful," he departed, telling us to be of good cheer and we should soon be honoured by an interview with his liege lord the Kaid. As he departed I responded to his compliments ceremoniously in English, devoting his liege lord to all the devils mentioned in the Talmud, Cabala, Apocalypse, and other works upon theology which treat of angels, devils, and like works of human ingenuity. Lutaif almost exploded, and even Swani, who knew a little expletive English, seemed amused.

After the interview we walked up the river and sat down underneath an algarroba tree, to watch thirty or forty people, men, children, and women, more or less carelessly veiled (for in this matter Berbers are much less strict than Arabs), all cracking almonds in a long open shed.

Almonds are one of the staple articles of trade in the Atlas Mountains. Long trains of mules, during the autumn, convey them to Morocco city; from thence they go to Mogador and are shipped off to Europe, " though what you Christians do with such a quantity of almonds we do not know." I took especial care not to enquire the price per bag, or ton, or box, they sell at; how many qualities there are; when they are ripe; how they are picked, sorted, or anything of that fatiguing nature, knowing how much I had disliked that kind of thing in books of travel I have read. Is not all that set forth in Consular Reports, in Blue Books, and the like, and who am I, by means of information got meanly at first hand, to " blackleg," so to speak, upon a British Consul, or to " springe cockle in his cleene corne " with unofficial and uncalled details which would not bring me in a cent?

Lutaif, and I, the Sherif from Sus, and the Persian, sat under a walnut grove for several hours, reading el Faredi, trying to learn the Berber tongue, talking of things and others, and I began at last to understand the lives passed by the Greek philosophers, who no doubt wandered up and down talking of things they did not understand, dozing beneath the trees, and weaving theories about First Causes, Atoms, Evolution, the Nature of the Gods, Schemes of Creation, and other mental exercises, at least as interesting as essays upon View of Frank Pledge (Visus de Francopledgii), Courts of Attachment, Swanimote, or any of the Judgments or the Entries of the Assizes of the Forest, either of Pickering or Lancaster.

Much did the Sherif enquire about the "Chimin di fer"
—how it was made, how fast it went and why, and if men
sat upon it (as he had heard) and worked it with their legs.
I told him all I knew (and more), and about bicycles and
women's rights; gave a brief digest of our common law and
current theology, tried to explain our parliamentary system,
the Stock Exchange, our commerce, glory, whisky, and the
like; told him of London's streets at night, our churches,
chapels, underground railways, tramways, telephones, electric light, and other trifles of our pomp and state which
might be interesting. His observations were few but pertinent, and certainly, upon occasions, hard to answer out of
hand. The bicycle, the Persian kindly corroborated; the
telephone, Lutaif explained; a railway underground was no
more wonderful than one upon the surface of the earth;
our parliament seemed foolish to him, and he could not
understand how one man could represent another, still less
ten thousand men; the Stock Exchange he summed up
briefly as a fraud, remarking, "How can they sell that
which they have not got?" but he reserved his chiefest comment for more simple things. "Whose are those wives," he
said, "who walk your streets? I thought you Christians were
monogamists. Do you not keep the law you follow then, or
are there not sufficient husbands for the women, or are your
women bad by nature; and how is it your Cadis tolerate
such things?" I told him, courteous reader, just what you
would have said in my position; confessed our faith in strict
monogamy, but said the flesh was weak; that even amongst
ourselves the Oulad-el-Haram were a large tribe; that all

worked towards perfection, and no doubt, some day, things would improve. He quite courteously rejoined, "Yes, yes, I see, you Christians are like us; that is, your faith is stronger than your deeds; well, even with us there are some men who read the Koran, and yet forget to follow all it says."

I dropped the conversation and urged the old Persian on to talk about himself; to tell of Persia, Turkey, the Sahara, and the Oasis of Tindoof, which he described by smoothing his hand upon the sand to show the flatness of it, and then said "Tindoof walou," that is, in Tindoof, nothing; though he confessed it is the mouth of Timbuctoo. In the same way Canada is said to have got its name from the two Spanish words "Aca" and "Nada," as signifying "there is nothing here." "Baris,"* he said was beautiful, the houris as the houris in paradise, but more expensive; and he detailed with great astonishment that once on entering a "cabinet" (presumably inodore) he had to pay a franc on coming out. And still he dwelt upon our hardness towards our co-religionists, and said, "Amongst you, look how many starve; here they may kill us, or throw us into prison, mutilate us, put out our eyes (these things are all against the Koran, as your women on the streets are clean against the teachings of your Book), but who can say a poor Mohammedan was ever known to starve? Look at me here," he said, "I came walking along the road carrying my broken fiddle, not cursing God, after the fashion of the Christians, for my bad luck,

* Paris is always pronounced "Baris" by Arabs. Vapor, a steamer, "Babor" and so on.

but praying at the Saint's tomb, leading the prayer in wayside mosques, calling to prayers when asked, maintained by all, till I arrived at this place and was received by the great Kaid with favour, given food and clothes and twenty dollars,* and fell into the disgrace in which you see me by my own act."

Though I knew of the Kaid's face being averted from him, and that he was about once more to take his fiddle up and walk, I had, out of politeness, to pretend my ignorance, and then he spoke.

"Yes, I was well received, and nightly the Kaid would send for me, and I discoursed to him of Persia, Turkey, and other countries which I had seen, and also of El Hind and China which I have not visited, but trust to visit before I call myself a traveller, and all went well and others envied me. Till, on a night, puffed up with pride, I let my tongue escape. Unruly member! (Here he drew it out and held it, like a slippery fish, between his finger and his thumb.) With it I told the Kaid that he might be as a great Sultan, but that I was greater, for was not I a Sultan of the mind? and since that time I have not stood once in the presence, and shortly I shall leave this place without a friend unless it be that you and this same noble Syrian, who speaks for you, can be counted in the number of the men who wish me well."

* Read perhaps " two dollars," for it is politic to exaggerate the munificence of the great. The politic man shall stand before kings, and they shall honour him.

He ceased and I determined later on to take the hint, and fell a-watching a small brown tree-creeper, not larger than a wren, which ran along the branches of a walnut-tree, and put its head upon one side and looked at us as it had been a fashionable philosopher, and no doubt just as wise. The grove of walnuts, sound of the running stream, at which a large grey squirrel sat and drank, noise of the wind amongst the leaves, the distant glimpses of the snow-capped hills, the long sustained notes of the negroes singing in the fields, set me a-thinking upon some not impossible Almighty Power, a harmony in things, a not altogether improbable design in nature, and, in fact, embarked me on the train of thought which the deep-thinking author of the libretto of " Giroflée et Girofla " has typified as thoughts about " de l'eau, de l'amour, des roses," etc., when the irruption of a scabby-headed boy summoned me back to earth.

He (the scabby-headed one) had heard I was a doctor of some repute, and had cured divers folks, as, indeed, I think I had, or at least medicined them out of self-conceit, and came to ask my help for the affliction which had come upon his head. Taking rather a hasty look, I almost had prescribed washing and an ointment with some butter mixed with sul-phur, when I bethought me what was expected of a doctor in the country where we lived. So, after thinking carefully for a little space, I said, " This requires thought, come to the tent in two hours' time (three hours were better), and I hope by then to have thought out, Inshallah, a fit medicine for your case." Duly interpreted into Berber this satisfied him, and he promised to comply. Our *tertulia** then dispersed, and we went back to sleep inside the tent, or resting on our rugs, to watch the constant passing stream of white or brown-dressed, bare-footed figures that, in Morocco, is always to be seen passing upon every road.

Considering that, in the Romans' time, Pliny, quoting from the lost journal of Suetonius Paulinus, informs us that the Atlas Mountains were full of elephants, and swarmed with wild beasts of every kind, who took refuge in the impenetrable forests with which those mountains then were clad, it is strange to find the Atlas of today so largely a treeless and a gameless land. True, that much later than the Roman times, the white wild cattle, now confined to Cadzow, Chillingham, and Lyme, roamed through the glades of Epping Forest, where today nothing more fearful than sandwich paper, or a broken bottle, makes afraid the explorer, and that wolves and wild boars were common till almost modern times with us. The destruction of the forests can be accounted for by the constant burning for pasturage, and with the forests much of the game would go, but why, besides large animals, there should be a dearth of birds, I do not know. With the exception of a hawk or two, pigeons and partridges, the lesser bustard, the little tree-creeper,

---

* *Tertulia* is the Spanish word for a gathering of people sitting talking about nothing, or important matters, for amusement. Thus after an hotel dinner at a watering place, there is always the " Rato de tertulia," *i.e.* the half hour of conversation in which the affairs of the world and one's acquaintances are discussed and settled. The word is also used for an evening party.

before referred to, a species of grey wagtails, and a reddish-brown sparrow, I hardly saw a bird. The ibises that follow cattle, the wild ducks, water birds, the greater bustard, herons and large hawks, all so common in the plains, are here conspicuous by their absence. I saw no storks,* and I believe they rarely come so high into the mountains as Kintafi, though in the plains no Arab hut so small as not to have its nest, with its two storks chattering all day, quite as persistently, and as far as I can see, to quite as little purpose as do members of the Imperial Parliament.

As regards animals—hyenas, foxes, jackals, wild boars, moufflons, porcupines, grey squirrels, small hares, and rabbits in considerable numbers, are to be found, with a few wild cats, and now and then a panther; lions are unknown, except in the great cedar forests of the Beni M'Gild to the north of Fez. In many parts of the plain country of Morocco the animals above referred to (except the lion), but with the addition of the gazelle, are numerous, but in the Atlas Mountains, at least about Kintafi, they are scarce and hard to find. In the plains sand grouse and several kinds of plovers are frequent, but I never saw them in the hills. Certainly both in the mountains and the plains all men have guns, and some of them shoot well; but though they shoot partridges, ducks, and pigeons in the plains, and in the mountains hunt the moufflon, and the wild boar and porcu-

---

* Much offence has often been given in Morocco by "sportsmen" firing at storks. The Arabs, who are not civilized people, do not understand killing anything you cannot eat. Besides, the stork, amongst birds, is the friend of man, as the porpoise is amongst fish.

pine, I never saw an Arab or a Berber fire at a small bird;
so that they cannot have been all destroyed, as is the case
in certain parts of Europe, by the efforts of the " Sontags
Jaeger " and his twenty-five mark gun. Curious idea that
known as " sport," and perhaps liable, as much as anything,
for the degradation of mankind. Witness the Roman show
of gladiators, the Spanish bull fight, and the English pheas-
ant butchery, where keepers wring the necks of dozens of
tame birds to swell the bag which has to figure in the newspapers, so that the astonished public recollect Lord A. or
Mr. B. is still alive.

One thing is certain in Morocco, and that is, that the
domestic animals and man understand one another better
than they do in any other country I have seen. It may be
that it is because in the East the animals, now become so
terribly mechanical in Europe, were first domesticated. It
may be that the peaceful life gives time for men and animals
to make each other out. It may be that the monstrous cart
horses, grey-hound-like thoroughbreds, ridiculously cropped
and docked hunters, and capitalistic looking carriage horses,
with the whole Noah's Ark of beef producing, milk secret-
ing, wool growing missing link between the animal creation
and the machine world, which we see in the fat " streaky
bacon " pigs, disgusting short-horns, and improved sheep
with backs like boxes, feel their degradation, and hate us
for imposing it upon them and their race. No one can
say it is because the intellect of biped and of quadruped is
nearer in degree than it is in Europe, for between the Arab

tribesmen and the English or Scottish countrymen, the balance certainly is not in favour of the northerner, if abstract power of mind apart from education is to be the test. What makes a flock of sheep follow an Arab, and have to be driven by a European, I do not know. Why, if an Arab buys a kid, in a few days it follows him about, I cannot tell, but there is nothing commoner than to see Arabs walking on the roads with curly black-woolled lambs and goats walking along beside them, as only dogs, of all the animals, will do with us.

Curiously enough, in South America, where animals of every kind are much more plentiful than in the East, they are not on the same familiar terms with man.

Who ever saw a lamb follow a Gaucho, Texan, Mexican, or any man of any of the countries where sheep abound in millions, such as Australia or New Zealand? I do not say that in Morocco animals are greatly better treated than with us, though on the whole I think perhaps the scale inclines against ourselves, but still there is community of feeling which I have never seen in any other land. Sometimes this same community makes the tragedy of animal life even more hard to understand, as when a man is followed by a lamb or kid up to the butcher, and stands and sees the thing which followed him so happily have its throat cut before his face, sees its eyes glaze, and its hot blood pour out upon the sand, pockets his half-dollar, and walks serenely home, after a pious exclamation about "One God"; as if God, either in one or three, could possibly be pleased to see one of his own created creatures so betray another only because it walked upon four legs. However, let "the One" (El Uahed) be pleased or not, no section of his clergy throughout Europe have said one word in favour of good treatment of their brother animals. Popes and Archbishops of Canterbury, of Paris, York, Toledo, and the rest, are dumber than dumb dogs, fearing to offend, fearing, it may be, that the animals have souls, or daring not to speak for fear of the stronger brethren; for when did priest, tub-thumper, bishop, Pope, or minister of any sect, take thought about the feelings of the brethren who are weak?

My scabby-headed patient turned up towards evening, and sat expectant in the door of the tent. After some thought, I told him to rise exactly one hour before the Feyzer, and get water from a stream (not from a well on any account), then put it in a vessel and pray exactly as the Muezzin called to prayer, then to walk backwards round an olive tree three times; every third day for twice nine days, to avoid all food cooked with Argan oil; finally, to wash his head well every day with soap, then to rub in butter mixed with sulphur, and then, if God so willed it, he would be well. A Seidlitz powder which he drank at once, and said there was "a spirit in the water,"* and nine Beecham's Pills to be taken alternately upon the third, the seventh, and the ninth day till they were finished, sent him away rejoicing, and laid my fame for ever as a first-class "tabib."

* Literally a Djin. The belief in the Djinûn seems to be a relic of Pantheism, or some older faith than that of Mohammed.

We now began to think how far our messenger had got, and were discussing how great the astonishment of the few Europeans in Morocco city would be on hearing of our fate, when the tent door was lifted and our messenger walked in, and silently laid the packet and the two dollars at my feet. Had a volcano opened upon the neighbouring hill I could not have been more surprised, and for a moment no one spoke, so much had we counted on the letters being well upon their way.

Swani first broke the silence with a string of imprecations on the unlucky messenger's female relations whom he defiled, gave up to Kaffirs, compared to hens, cows, goats, and finished up by telling the poor man he evidently was born of a family the women of which were shameless, veilless, and as hideous in their persons as their characters were vile. The poor man sat quite patiently, and then replied it was no use to curse him, his heart had failed, and that he feared if he were found out the Kaid would kill him, burn his house, and throw his children into prison to rot and die. Though I was much annoyed, I was sorry for him, as one is always sorry for all those whose hearts fail at the wrong minute, and I was touched that he had brought me the two dollars back. Most likely in his life he had never seen himself at the head of so much capital, and it would have been easy for him to throw away the packet and not return. Therefore I handed him a dollar, and remarked, those who have families should not engage in enterprises such as these; God loves stout-hearted men, but perhaps loves quite as much fathers who love their children, but, children or no children, we are all in his hand. This, though they had heard it a million times, seemed to console all present, and the messenger slunk from the tent.ashamed, but happy, having been paid on a scale he thought was lavish for a mere twenty-two hours' walk.* Certainly when he went we were cast down, for it appeared impossible to get a letter safely conveyed, so we agreed next morning to saddle up our animals and see if the Kaid would allow us to return, thinking, perhaps, his injunctions only lay upon the road towards the Sus. This settled, Lutaif and I walked long backwards and forwards on the Maidan, in the clear moonlight, and heard the long-drawn, quavering notes of a wild song like a Malagueña rise in the still night air, wayward and strange in interval, sung in a high falsetto voice, and yet enthralling and penetrating to the marrow of the bones; once heard, haunting one's memory for ever afterwards, and still almost impossible to catch; but it recalls Kintafi to me as I write, just as the scent of fresh-cut oranges brings me back to Paraguay, so that perhaps perfumes and sounds are after all the most stable of the illusions amongst which we live.

* After he had gone a doubt occurred to me whether he had left the place at all, and had not merely hidden himself, and came back to amuse us with a comedy. But I consoled myself by thinking that even if this were so, I had paid more money to see worse acting in a theatre.

Chapter VIII

JUST about daylight we began to load our beasts, looking anxiously the while to see if any notice of our proceedings was taken from the castle walls. No one stirred, and hungry, without provisions for the road, our animals half-starved but lightly laden—for the greatest weight we had in coming had been food and barley—we prepared to start.

In the other tents the people made no sign, it was so early that neither the slaves were in the fields, nor yet the prisoners come up out of their living tomb, and still I thought it would be prudent before leaving to send Mohammed el Hosein to say that we were going, for to escape unseen was quite impossible, and even if we had slipped off unseen, once the alarm was given we should have been overtaken and brought back at once. We had not long to wait; Mohammed el Hosein soon came back crestfallen, the postern door was swung wide open and the Chamberlain emerged, followed by several tribesmen all ostentatiously carrying long guns. Although it was so early he was dressed, as at all times, in most spotless clothes, and walked across the Maidan with as near an approach to haste as I had ever seen him make. Arrived at where we stood, he saluted us quite ceremoniously, and asked where we were going, to which I answered, "Back to Mogador." On this he said, "The Kaid bids me to tell you not to go today as he could never think of letting Europeans go without an audience, but most unfortunately his wound pains him this morning, and besides that, now you are known as Christians, he would not let you wander through the hill passes without an escort, therefore he bids me tell you to unload and wait." For a moment I thought, "If we go on he will not dare to stop us," and taking my bridle in my hand, prepared to mount, when the armed followers drew near, handling their guns, as if to shoot a Christian would have been great sport. The Chamberlain said a few words in Shillah, which having been interpreted, said if we insisted upon going he must see his master's orders carried out. Seeing that the Kaid was resolved we should not go, I gave my horse to Swani and went into the tent. The Chamberlain came after me, and standing in the door told me most civilly that he had done what he was told to do; as he had done it in the most well-bred way, with every consideration for my feelings and without a trace of swagger, I thought the moment had arrived to talk and understand each other if we could. The Chamberlain, Sidi* Mohammed, was a well-favoured, "coffee and skim-milk " coloured man; portly, of course, as

* Sidi originally meant Lord in Arabic. Today, in Morocco at any rate, it has, like Esquire, fallen from its proud estate, and now about answers to "Mr." It is used in addressing anyone who has good clothes, the chief class distinction in Morocco, where all are socially equal to an extent unknown in Europe, except in Spain. Sidi was the word from which the Cid took his title.

became his office, honest as officers of great men go, well-dressed and courteous; in fact, a sort of Eastern Malvolio, with the addition of some sense.

I laid before him my two chief complaints, which I said I had no wish to bother him about, but that it seemed the best thing I could do was to marry a maiden or two belonging to the tribe and set up house, as there seemed little chance of ever moving from the place. However, in the meantime, should the Kaid consent to let us go, I did not want to walk back to the coast, and my horse and the other animals were growing weaker every day for want of food. Without preamble, therefore, I promised Sidi Mohammed a handsome present when I went if he would see that the man who kept the corn gave a sufficient quantity every day and did not sell it or keep it back as he had done for the past week. Sidi Mohammed expressed astonishment at such behaviour, and perhaps felt it, as no doubt the Kaid had ordered our horses to be fed, and promised to see about the matter instantly and put things right.

My next complaint was that there were five of us all in one small tent, and that such crowding was neither comfortable nor seemly, either for a Christian " caballer " or for a Moorish gentleman, which rank my clothes and following entitled me to take. The promise of the present smoothed the way, and Sidi Mohammed said he would take upon himself to give permission to pitch another tent. This being done, and the men, the saddles, harness, and saddle-cloths transferred to a smaller tent, we had our own swept out and aired; a new drain cut to carry off the water, and stones arranged (which looked exactly like an Arab grave), to place our rugs upon, and keep them off the damp. For the first time for a week Lutaif and I were comfortable, washed in a tin basin, changed our clothes, and sitting in the sun at the open door of the tent drank tea and smoked, planning the while to get another man to take the letters which the faithless or heartless messenger had brought back the night before. Swani and Mohammed el Hosein went to the river and washed their clothes, and even Ali, who had nothing except what he wore, borrowed an old Djellaba and, standing in the river, stamped upon his rags.

Our friend the Persian came and sat with us, condoling on our having been prevented starting, but saying I had taken the right way with Si-Mohammed, and that he was glad we had got another tent.

As we sat talking, a Jew pedlar arrived bringing two laden mules. The Persian said he might by chance have some tobacco, and being out of it I sent and asked the Jew to come and talk. He came, and thinking I was a Moor began to offer all his goods, henna, and looking-glasses, needles and cotton, scissors from Germany, knives made in Spain, and cotton cloths (well-sized) from Manchester. I let him talk, but when he saw that every now and then I missed some words and had to ask an explanation from Lutaif, who answered me in English, he began to stare, and at last said in stumbling Spanish, "Are you not a Moor?" "No," I replied. He said, "What, then, you cannot be a

Jew?" On hearing that I was a Christian,* his amazement knew no bounds. " Christian," he said, " and dressed like a Moor, camped in the middle of the Atlas, how ever came you here?" When I informed him I had passed as a Sherif, he roared with laughter, and said he would have given all his mules' load to see the people come and kiss my clothes.

It seemed he lived in Agadir-Ighir,† and traded through the Atlas, as he informed me many of the poorer Jews do, selling their goods, and buying wool and goat-skins to take back. He had tobacco from Algeria of a villainous quality, strong, black and common, and done up in gaudy-coloured packets, adorned, one with a picture of a lady dancing " le chahut," another of apocryphal-looking Arabs resting in an Oasis; the third displayed a little French soldier running his bayonet through a picture of Bismarck, and underneath the legend, in Italian, " Furia Francese," and to make all sure the Regie mark. I bought all three, which sold him out, and all my men were gratified with about half-a-pound apiece. They said it was the best tobacco they had ever smoked, but I think that tobacco was to them as it was to a Scotch gáme-keeper to whom I gave, when a boy fresh from school, a packet of Honey Dew which I had bought in London, and who said upon my asking him if it was good, " Ye ken, Sir, if she burns she is good tobacco, and if she willna burn, then she's nae good." The Persian too participated in the

---

* Christian is official in Morocco. If you are not a Mohammedan or a Jew, you are a Christian.

† Ighir means a castle or fortified place, in Shillah.

tobacco, being reduced to smoking Kief.* Under the influence of the Algerian tobacco, which, to make himself intelligible to me, he characterized as being " bon besaf," he got back to his wanderings up and down the world.

Ifrikia, as he called Africa, he thought the most savage and abominable portion of the earth. Even the Kurds, whom he knew well, he thought were not so fierce as were the Arabs of the Wad Nun. The poor man, an ardent believer in Mohammedanism, though not a bigot, and at times gaining his livelihood by discoursing on Mohammed and the Koran, whilst traveling in Wad Nun upon the road to Timbuctoo, which as he said he did not reach, there being " too much powder on the road," was frequently in peril of his life, being taken for an unbeliever, being himself a Sufi, and the Moors all members of the sect into which orthodox Mohammedans are grouped. The poor old Ajemi† it appeared on one occasion was surrounded by a band of Arabs who held their daggers to his throat, and put their guns up to his head until he, losing patience, knelt upon the sand, said " Bismillah, kill me in God's name," reciting the confession of his faith in a loud voice. However, Allah, he said,

---

* Kief is hemp mixed with some other herbs and cut up fine and smoked in little pipes about the size of those used for smoking opium. It does not produce torpor as the Turkish Hashish does, but if too much indulged in destroys the health and gives a curious livid look and glaziness of the eyes to the habitual smoker. Taken in moderation it has the sustaining qualities of the Coca of the Andes and the Betel Root of Java and the Far East. A slight headache is all that I have ever experienced on smoking it.

† Ajemi is the Arab term applied to any foreigner as distinct from themselves. As at the inception of Islam the Persians were almost the first foreign nation they met, the term has become used exclusively to designate them.

had spared him, for after taking all his money, and almost all his clothes, the Arabs had let him go, and cautioned him to walk with God and not return to the Wad Nun again. This he was confident he would not do, preferring even Franguestan and its peculiar ways to the companionship of such evil-begotten men as those. I like to think of him, friendless and all alone, kneeling upon the sand, surrounded by a crowd of horsemen, ready, although not wishing, to be killed, and wonder if he thought about the irony of things, that he, an ardent votary of his religion, was to be put to death for heresy. At times when thinking upon other people's travels (always so much more interesting than any of my own), it comes before me how that, in desert places, mountain passes and the like, so many men must have been killed, and met their fate heroically, the situation so to speak thrown away, with no one there to see, record, to write about it, as if the poor, forlorn and wasted heroes were no more worth a thought than the fat man of business who snorts his life out on a feather-bed between a medicine bottle and a mumbling priest. So the old Persian left us to make his preparations for an early start next day, hoping to reach a saint's tomb of great sanctity on the hill path which leads from Kintafi to Tamasluoght, but is only to be passed on foot. He said he was tired of this wild part of Africa, and would make his way to Mogador, thence to Tangier, return to Persia and push on to China, which he hoped to visit ere he died. Considering that he had little money, for I expect the twenty dollars of the Kaid were for the most part quite

apocryphal, and that the journey, made as he would make
it, would probably take years, he did not seem too excited, or
as much so as a man who thinks his things have been put
into the wrong luggage van at Charing Cross.

People who write about the progress of the world, the
wealth of nations, of economic laws, and subjects of that
kind, requiring rather stronger imaginative powers than
reason, logic, or than common-sense, are apt to take it as a
well-established fact that before railways were invented
people, especially poor people, travelled but little, and generally never moved far from the places where they were
born. This may have been so in the last two hundred years,
although I doubt it, but certainly during the Middle Ages
they must have travelled much. Leaving the pilgrimages out
of account (and they, of course, brought every European
nation into contact), I take it that many roved about, as they
still do in Eastern lands. People, no doubt, had no facilities
for travelling for mere amusement's sake; but if we read
any old book of travels, how often does the writer meet
a countryman, a student, minstrel, soldier, or wandering
artisan in countries far away.

So, when the Persian went, we strolled out for a walk,
followed the river for a mile or two, and found it full of
fish; but the whole time we sojourned at Kintafi we saw no
one fishing either with rod or net. The people whom we
met were all well armed; and when they met us, kissed our
clothes, taking us for Arabs of rank upon a visit to the Kaid.

It always pleased me to see two Arabs or two Berbers

meet, embrace each other, kiss each other's shoulder, ask respectively, How is your house? ("Dar-de-alic"), for to enquire after the health of even a brother's wife would be indecent; and then, the ceremony over, sit down to talk and strive with might and main to cheat each other, after the fashion in which Englishmen proceed in the same case.

Seated beneath a cliff, our feet just dangling in the stream to cool, smoking the vile Algerian tobacco, Lutaif began to tell me of his life in Syria, described his father's house, a great, gaunt place with a long chamber in the middle, given up to winding silk; spoke of the undying enmity between the Turk, the Druse, the Maronite, and the Old Catholics, of which sect he was a member; leaving on my mind the feeling that the Lebanon for a residence must be as undesirable as was Scotland in the old wicked days, when they burnt witches, and the narrow-minded clergy made the land a hell. One thing particularly struck me when he said, upon a walk, if we had been in Syria, dressed as we were in clothes which marked us for Mohammedans, and had we met four or five Christians, they would have either insulted or attacked us; and, of course, the same held good for Christians who on a walk met Turks. Remembering this happy state of things, and having from his youth looked upon every Mohammedan as a sworn enemy, when he first came to Morocco, knowing the people were fanatical Mohammedans, he passed his life in dread. Once in Tangier, not thinking what he did, or of the peril that he naturally incurred, he took a country walk. He started from the town dressed as a European, carrying a silver-headed stick, and several oranges in a brown paper bag to eat upon the way. After a mile or two, he took an orange out of his bag and, sitting down, was just about to eat when to his horror, on the sandy road, what did he see but five or six well-armed young men come, as he said, dancing like devils up the road and brandishing their knives. He called upon his God and closed his eyes, being quite sure that his last hour was come. Then to his great surprise the men stopped dancing, sheathed their knives and after saluting him respectfully sat down, several yards away, without a word. At last one asked him humbly for an orange, and Lutaif took the whole bag and was about to entreat them to take all his oranges, his clothes, his money, everything, but to spare his life. To his amazement, the man took an orange from the bag, divided it into five portions, one of which he handed to each of the young men, and handed back the bag. The exiguous portions of the orange discussed, the spokesman asked him not to point the silver-headed stick their way, for it appeared they had got into their minds it was some kind of gun, which, if Lutaif discharged it, would destroy them all. He promised faithfully, and the wayfarers went upon their way, leaving Lutaif as frightened as themselves. No doubt when first he saw them they were exercising, skipping about like fawns, in the sheer joy of life, but as they came upon him suddenly, their sandalled feet making no noise upon the sandy road, for a Syrian the vision must have been horrible enough.

When he wrote home and told his friends what had occurred to him, and how a Christian was regarded (near Tangier) with respect and awe, the answer that he got was curious. Of course they thought he was telling lies, but his best friend admonished him it was bad taste to jest about Mohammedans; for, though no doubt they were bad neighbours, no one in Syria could call them cowards. In fact, the friend appeared to me to be like every Eastern Christian I have met, quick to run down the Turks, to fight with them, hating them bitterly at home; but yet if a stranger slighted them abroad, quick to resent the slight, say they were brave, and that they erred through wicked counsellors and not from lack of heart.

All the above he told me, and plenty more, with the inimitable charm that Easterns have in story-telling, compared to which even Guy de Maupassant, Chaucer, Boccaccio, Balzac, or Fielding fall immeasurably behind. The doubtful author of the Celestina* and Cervantes, perhaps, come nearer; but then they, being Spaniards, were more nearly in communion with the East.

As we returned, tired and half hoping that there might be news, we learned a " rekass " had just arrived bringing despatches from the Sultan to the Kaid. Though we knew well there had not been time for the Kaid to get an answer to the letter he had sent about ourselves, yet outside news

---

* Mr. Fitzmaurice Kelly is confident that Rojas threw off the Celestina in a fortnight's holiday, but if so, I cannot help wondering why he threw off no more, as writers, ancient or modern, seldom know the force of the American adage, "When you strike ile, stop boring."

was valuable, and, in fact, had I chanced to come upon a
copy of the Rock, I think I could have read it, advertise-
ments and all. In half an hour or so the "rekass" strolled
past our tent, and I invited him to come and have some
tea. Out of respect he sat outside the tent, saluted us, and
remained waiting to be itnerrogated. He was a tall, lean,
teak-complexioned man, in face resembling a Maori god
stuck up outside a Pah; vacant and glassy eyed and at first
sight a kieffi, that is a kief smoker, thick lipped and with
uncertain speech as if the tongue was (like the tongues of
Bourbons) too large for his mouth, also a symptom of too
much kief smoking; legs like a broncho's from the Bad
Lands, a mule's, or a bagual's from the stony deserts of
Patagonia; feet rather large, with the toes so flexible that the
whole member seemed to quiver as he walked. For clothes
he had a single white garment like a night-shirt (long freed
from all the tyranny of soap), hanging down almost to the
ankles, girt round the waist with a string of camel's hair. He
went bare-headed and had a cord of camel's hair bound
round his temples, with a long lock, at least eight inches
long, hanging from the top of his bare shaven head beside
his ear. Though he had walked incessantly for the last seven
days, sleeping an hour or two with a piece of burning match
tied to his toe to wake him as it burned away, he strolled
about, or sitting drank his tea, taking a cup now and again,
which Ali or Swani passed to him out of the tent. He said as
long as he had kief he never wanted food, but munched a
bit of bread occasionally, drank at every stream, and trotted

on day after day, just like a camel, for, as he told us, he was
born to run. Withal no fool, and pious, praying now and
then whenever he passed a saint's tomb and felt wearied
with the way. Just such a man as you may see amongst the
cholos of the sierras of Peru, with the difference that the
cholo takes coca instead of kief, and is in general a short,
squat, ugly fellow, whereas our kieffi stood over six feet
high, straight as a pike-staff, and was intelligent after his
fashion, could read and write, and no doubt knew as much
theology as was required from a right-thinking man.

For impedimenta he had a little bag in which he kept his
kief, his matches, pipe, and the small store of money which
it was possible he had. In one hand he carried a stout
quarter-staff full five feet long, which all " rekasses " use to
walk with, try the depth of water in crossing streams,
defend themselves, and ease their backs by passing it behind
them through their two arms, and resting on it as they trot
along.

His news was brief but bloody. " Our Lord the Sultan is
camped in Tedla.* He is indeed a king, fifty-one heads cut
off, two tribes quite eaten up, three hundred of the Kaffirs
wounded! O what a joy it was to see the ' maquina,' the
Christian devil gun, which fires all day, play on the enemies
of our Lord the king. Praise be to Allah who alone giveth
victory." Which being interpreted meant that the Sultan
had gone under pretence of peace to Tedla; had by the

* A district a little north-east of Morocco city in which the tribes are in
constant rebellion.

advice of the Grand Vizier Ba Ahmed attacked them;
butchered as many as he could, and probably sent a few
hundred to die in gaol. The selfsame fate overtook the
Rahamna tribe close to Morocco city. They fought a year
with varying success, but at the last were decimated,
butchered in hundreds, and their power destroyed.

The Grand Vizier Ba Ahmed, if all reports be true, is a
bad counsellor for the young Sultan, Mulai Abdul Aziz.
But be this as it may—for some who know the country say
that the Grand Vizier, being a Moor, knows how to rule
his countrymen—Sidi Ahmed ben Musa, usually called Ba
Ahmed (Father Ahmed), is an ambitious and most powerful man, holding the Sultan in a sort of tutelage, and piling
up a fortune by his exactions, which report says he has
invested in safe securities abroad.

The father of the Sultan, Mulai el Hassan, who died or
was poisoned some four years ago, was a remarkable per-
sonality, and perhaps one of the last Oriental potentates of
the old school. Standing about six feet three inches in his
slippers, he was dark in face, having, though a descendant
of Mohammed, some negro blood; a perfect horseman, shot,
and skilled in swordsmanship; though educated in all the
learning of the Moors, he yet was tolerant of Christians,
kind to Jews, and much more liberal in regard to new ideas
than is his son, that is to say, if it is not Ba Ahmed who
directs his policy. Mulai el Hassan was what is called a
" riding Sultan," that is a warrior, always on horseback, and
passing all his life either in journeys between his various

capitals, or on long expeditions to reduce refractory tribes. His fine white horse has been described by almost every embassy for the past ten years that went to Fez, for from his back the Sultan used to receive ambassadors, who bound in their hats, hosen, coats, swords, tight boots, and dignity, and forced to stand in a hot sun, on foot, must have presented a very lamentable sight.

On the white horse's back the Sultan almost died, for one who saw him shortly before his death was standing in a street in the outskirts of Marakesh when the Sultan passed, having been sixteen hours on horseback in the rain, and looking like a corpse. Next day he died so suddenly that some thought he had been poisoned, but others think worn out with care and trouble, long journeys, and all the burden of a ruler's life. All those who knew him say that his manners were most courteous, kind, and dignified, and that through all his life none of his servants ever heard him raise his voice, even in battle or when he ordered some unlucky man to death, above its ordinary pitch.

His clothes were spotless white; but made in the fashion of those worn by an ordinary tribesman, only of finer stuff. Colours he never wore, or jewellery, except a silver ring with a large diamond, and which when once an individual, whose name I forbear to mention, asked him for it for a keepsake, he half drew off (for usually he gave all that was asked for); but replaced and said with a quiet smile, " No, I will keep it, but you can have its value in money if you choose." His clothes he never wore more than a day, and then his servants claimed them as perquisities; so that his wardrobe must have been pretty extensive even for a king. Upon a journey he carried almost all he had, packed upon camels, and, being troubled with insomnia at times, would say, " Bring me the telescope the Belgian Minister gave me ten years ago," or " the watch the Queen of England sent me," and the unlucky man to whom he spoke had to produce the thing, if he unpacked a hundred camels in the search.

The taxes he used to collect in person with an army, so that his camp was like a town of canvas, and yet the order of his own tents so great and his men so skilled in pitching them, that at a halt they used to rise like magic from the ground.

Wives, and that sort of thing, he had about three hundred, and was much addicted to their company, and some of them accompanied him on all the journeys which he made. His son, the present Sultan, was born of a Circassian, white, and report said beautiful and educated; but she transmitted little beauty and less education to her son, who is a rather heavy youth of about twenty, not well instructed, and completely in the hands of his Vizier, Ba Ahmed, who, by exactions, cruelties, and bloodshed, has made his master's name detested all through the land. Still a strong man, and no doubt in such countries as Morocco, when a Sultan dies a strong man is required, for the tribes usually rise in rebellion, kill their Kaids, burn down their castles, and a recognized period of anarchy takes place, known as El Siba by the natives, and of which they all take full advantage.

That the Vizier was a man of readiness and resource is shown by the way be foiled the expedition of the yacht *Tourmaline,** by means of which a syndicate in London endeavoured to produce a footing in the Sus. Thinking the kieffi would be an excellent man to take our letters, we sounded him, but in a moment he became mysterious, said he must sleep, would think about it, and though we often saw him subsequently strolling about, he never visited our tent again.

After the kieffi went, the Persian came to say good-bye, and sat long talking about Montenegro, where the people are all brave, and to his astonishment, for they are Franks, the women virtuous. He tells us that their enmity with the Arnauts is constant, and this he illustrated by linking his two forefingers to indicate a fight. " Brave men," he said, " who, when they draw the sword, never go back, and a fine country, but lacking the true faith." Then came the leave-taking, and I presented him with a dollar, which he has magnified in talking of it to at least twenty, and he, after a few well-chosen, dignified sentences of thanks, pressed his open palm against my hand, and then pressed it upon his heart, saying again, " Mesquin, may you have patience, and may Allah not open to you the Sultan's gate !" And so he took the road, shouldering his sack of " possibles," and in his hand a staff, and carrying, God knows why, a wooden board, and in a little faded away on the hill track, out of

* For the yacht *Tourmaline* see Appendix.

my sight and life. Vaya con Dios, I never knew his name, for he was not a man given to descending to particulars of such a kind, and it is rank ill manners to ask an Oriental what his name is, the fiction being that he is so well known, to ask would be impertinent. It may be that he may cast up some day across my path, for he is always on the march; but if he does not, in many an Eastern khan and fondak men will know of me, not by my name, for that he never knew, but as the Frankish stranger whom he met a prisoner in the Atlas, and who gave him gold and more gold, so that he had to buy a sack to carry it away. And at the saints' tombs, and in mosques, there he will pray for me (at least he said so), and I shall know that what he says will not be said in vain, for has not Sidna Mohammed himself averred that "the prayers of a stranger are always heard by God."*

So, sadly, as if we had lost one we had known from youth, Lutaif and I wandered along the river; and by a stony beach, under some oleander bushes, came on a little tea-party, all seated on the ground. A little pleases Arabs, who in a measure are like children, easily pleased, and passing easily from good temper into rage, and nothing gives them greater satisfaction than when a stranger comes and joins their pleasure or their meals. So we advanced, and found they were three Sherifs from Taseruelt; Sherifs, but practically beggars, though white men of pure Arab blood and

---

* Swani, so that the *odium theologium* might not be wanting, hoped that the Persian would be killed on the road, for, said he, these Persian heretics are worse than Christians. The Persians are, as is well known, unorthodox, and who does not prefer an infidel to an unorthodox believer?

race. One was a little thin and wizened man, with hardly any beard, his clothes quite clean, but washed into holes by frequent soaping and thumping against the stones of streams. Quick, taciturn, and most intelligent, a hunter and, I think, an acrobat, and wearing round his head a yellow cloth gun-case twisted like a turban, which, with his meagre features and pale face, gave him an air as of a dwarf ghost or spectre, as he sat smoking kief. The other two were fine young men, but poorly dressed, and perhaps got their living by praying, or by writing charms, for all could read and write, and neither of the three seemed ever to have done any of that same honest toil which so much ennobles man. Placed on the sand before them stood a small brass tray, and on it three small glass tumblers and a tin teapot of the conical pattern which Germany supplies. Dried figs and walnuts were on another tray, and all were smoking kief. Close to them, on a little patch of grass, fed a black curly lamb, which I supposed they had reared and brought with them from Taseruelt; but they assured me it was given to them only two days ago, and now followed them like a dog. I asked if they intended to dispose of it, and they said no, they would teach it to do tricks, and gain much money by its antics, and as we spoke it walked up to the tray, took up a fig, smelt at it, but thought it unfit to eat, and then, after skipping about a bit, came back and went to sleep with its head resting upon its special owner's feet.

We squatted down beside the three Sherifs and became friends at once, drank endless cups of tea as sweet as syrup, ate figs and walnuts, talked of Europe and of Taseruelt, and, I think, never in my life did I enjoy an afternoon so thoroughly. They asked no questions, thinking it apparently not strange we should be there dressed as Mohammedans, and I almost unable to speak Arabic, as if, for example, a Chinese dressed as an English country gentleman should stumble in upon a gang of haymakers in Rutlandshire, and sit down and drink beer. Much did they tell of the Wad Nun, and of the desert horses, known as "wind drinkers," on which men hunt the ostrich, feeding them well on dates and camel's milk, and flying through the sands after the ostriches in the same manner that the Pehuelches hunt their ostriches in Patagonia, save that the Arabs throw a club instead of the ostrich "bolas" which the Pehuelches use. In both countries the tactics are the same, the huntsmen spreading out like a fan, striving to join their ranks and get the birds into a circle, or to drive them into a marsh, edge of a stream, or some place out of which they cannot run.

The little dwarf Sherif got up and showed me how an ostrich ran, waving his arms and craning out his neck in a way which would have made his fortune on the stage. It then appeared he had been a moufflon hunter, and he told how they can jump down precipices alighting on their horns, how shy they are; and here he worked his nose about to show the way they snuffed the wind when danger was about, so that he looked more like a moufflon than the very beast itself. His friends smiled gravely, and said Allah had given their comrade excellent gifts, and one was to be able to imitate all beasts, and another was to run all day and never feel fatigue.

On hearing this I mentally resolved he should run to Morocco city with our letters, starting that very night; but mentioned nothing of my purpose, intending to leave Swani to arrange it by himself. We thanked our entertainers, gave them some of our Algerian tobacco, which they prized highly, and the deputation then withdrew. As I looked back they had not moved, but the black curly lamb had gone back to the grass, and they, beneath the oleanders, seated on the stones, still sat smoking happily, before their little tray, as if the world belonged to them, as after all it did.

Just before nightfall Swani brought the small Sherif ostensibly for medicine to our tent. When asked to carry letters he said yes, that he was poor and wanted to buy clothes for winter, and would go at once, and his companions and the lamb could meet him somewhere near the coast. I asked if he could run, and he replied " like an Oudad,"* and by that name we knew him ever since. Five dollars was our bargain, two in the hand and three upon arriving at the missionary's house. He asked no questions save the position of the missionary's house, took the two dollars and the packet and a note asking the missionary to pay him three dollars when the letters came to hand, thanked us, and said " Your letters shall arrive," walked quickly off, and disappeared into the night. On the eve-

* Oudad, the moufflon.

ning of the third day from that on which he went, a dusty little man knocked at the missionary's door more than a hundred miles away, handed a packet in, and waited whilst the note he brought was read, got his three dollars and an extra one for speed, and when the missionary, who went for a moment into his house to read the letters returned to question him, he was already gone. So the Oudad after the Persian flashed across my path, or I intruded upon theirs, we talked, made friends, separated, and shall never meet again; but the impression that they made was much more vivid than that caused by worthy friends whom one meets every day and differentiates but by the checks upon their shooting jackets.

Determining to leave no stone untried, Lutaif, who fancied himself on his epistolary style, said he would write a letter to the Kaid to ask for an interview. About an hour he spent upon the task, lying upon his stomach in the tent, and writing on a large flimsy sheet of Spanish notepaper with a small pencil end; but after so much trouble he produced a gem, crammed full of compliments, in such high Arabic that he thought none but the Taleb would decipher it, and written as beautifully clear as Arab copperplate:

"To the most happy and exemplary, the most fortunate and honourable, the Kaid Si Taleb Mohammed el Kintafi.

"May God's peace and blessing be upon you whilst day lasts and time endures. O Kaid, thou art the wielder of the sword and pen. Fate and a love of travel have led us

to your happy and well-governed land, and you have generously received and entertained us, extending to us all
the hospitality of your thrice blessed house. May God establish it for ever, and may the hand of no man be ever higher
than your hand. But, mighty prince, we fear to trespass
too long upon your kindness, though we know your hand
is never tired of shedding blessings upon all. Therefore,
we wish to see your face and thank you for your hospitality, so that on our return on talking of you we can say
this was a man. May Allah bless and keep you, and at the
last may Sidna Mohammed welcome you upon your entrance into Paradise. Deign, therefore, to accord an hour
tomorrow on which to speak with you."

This missive, read aloud, evoked great admiration both
from Swani and Mohammed el Hosein, and they declared
it certainly would have a good effect. So Swani, dressed
in a clean white burnous, which Lutaif had with him in
his saddle-bags, and with a pair of my new yellow slippers,
went off, and with much ceremony handed the letter to
the keeper of the gate. Knowing the respect the Moors
attach to letters, and the astonishment they show if any
Christian can write their characters, I thought perhaps the
letter might bring an interview; but thinking of the happy
afternoon I had spent with the simple fakirs upon the
stony beach, did not care much, knowing the happy hours
that a man passes in his life are few, and of more value
than much gold or all the jewels of the Apocalypse.

Chapter IX

THOUGH not so sanguine as Lutaif, as to the emollient powers of his epistle, I was pleased to find that for the first time, next morning, we received ample supplies of food, baskets of grapes and oranges, and for the first time people spoke to us without an air of breaking some command.

During the morning a miserable bundle of rags arrived and stood before our tent, asking in broken Arabic if we were the Christians, and on being answered in the affirmative broke out into French. It appeared he was a French deserter from Algeria, having deserted in Ain Sefra,* walked to Figig, and pretended to turn Mohammedan, he came by Tafilet, and was about to make his way down to the coast. This, as he said, was his itinerary, but why he should have come round by Tafilet, he did not explain. He certainly was not a personable man; a weasel-faced, pale, and fair-haired Parisian " voyou," thin, active, and half-starved, foot-sore and weary, dressed in rags, and speaking a jargon of bad Arabic, compared to which that

* Ain Sefra is, or was a few months ago, the terminus of the French railway system in the Western Sahara. I should not be surprised that it was pushed on close to Figig by now, for the French in matters of this kind are not hampered with conscience, Nonconformist or otherwise.

spoken by the Persian and myself became as the language of the Khoreish, or the best literal Arabic which Cairo boasts. He told us that he slept in the mosques, making the profession of his faith if there was any doubt about him before going in; this with a wink, and " Sont-ils bêtes, ces Arabes, à la fin!" After he had eaten and smoked, he said that it was common in Algeria for soldiers to desert, adopt the Arab dress, and make their way into Morocco; some reached the coast, but many disappeared, murdered by the tribesmen or the villagers upon the way. Withal a merry knave, relating how he had served in a Spahi regi-ment during the war in Madagascar, and that the Arab troopers, when the war broke out, talked of the war with Madame Casba,* and thought she was Sultana of some island, who was fighting with the French. Although he had no arms or money, he did not seem afraid, but trusted to arrive in Mogador or Saffi in a few weeks' time. We went to bathe and left him smoking under a tree with Swani, talking a mixture of Arabic and French: on our return in half an hour, thinking to see him still before the tent, and make him tell us what he had seen in his long tramp, we found that, without a word to anyone, he had slunk mys-teriously away.

Once in Morocco city I met three Englishmen dressed in the red baize rags which form the uniform of the Sheri-fian troops. Where they came from they did not say, but

---

* " Uaheda Sultana Madame Casba."

wanted money to buy magia* and tobacco; I gave them
something, and on receiving it with not too laboured thanks,
they too mixed with the crowd in one of the bazaars and
disappeared.

In the crowded Kaiseriehs of the towns, and in the endless processions of noiseless-footed people on the roads,
nothing is more surprising than the way in which odd
characters come to the surface for a moment (like a fish
rising), and then sink back again into the depths from
which they rose.

On mules and donkeys, on horseback and on foot, beggars, or travelling well attended, Berbers and Arabs, Jews,
Negroes, Haratin, men from the Sahara, and from the
mountains of the Riff, Syrians, and Levantines, outcast
Europeans, and an occasional Hindu, with Turks and
Greeks, and people from the utmost regions of the Oriental
world, they all are there, and always on the move, travelling about as if some not too swiftly circulating quicksilver
ran in their veins; whither they go or why, whence come
from, and what urges them to wander up and down, is
to me inexplicable, and forms one of the many of the un-
fathomed and unfathomable problems of the East. Not that
I mean the various passengers whom I have named bulk
largely in the population of Morocco, but they are there,
and every now and then one feels how all the Oriental
world is linked together by nomadic habits, from Bagdad

---

* Magia is a spirit made by the Jews in Morocco; it is sometimes made of
grapes, sometimes of figs, and again of dates. The magia made of dates is less
lethal than that from figs or grapes. It is of a milky colour and very strong.

to Wad Nun, and from Shiraz to the oases of the Sahara.

In Morocco the prevailing tone is greyish white; men's clothes, and houses, towns, bushes, tall umbelliferæ, nodding like ghosts in autumn, all are white; white sands upon the shore, and in the Sahara, and over all a white and saddening light, as if the sun was tired with shining down for ever on the unchanging life. In no part of Morocco I have visited does the phrase " gorgeous East " have the least meaning, and this is always noted by the wandering Easterns, who find the country dull and lacking colour compared to Asia, or as the Arabs call it, " Blad Es Shark."*

Almost all day on the Maidan behind our tent football went on (called in Arabic El Cora), and every one joined in, middle-aged men, slaves, and the various hangers-on about the place, the Kaid's sons playing furiously and whilst the game went on they were not more respected, and received as full a share of kicks, shoves, trips, and pushes as did all the rest. The ball they used was little larger than a pomegranate; no rules seemed to be observed, for everybody pushed, shoved, bit, scratched, and kicked as it seemed best to him, and as they had no goals, but played simply to drive the other players back, the play was wild, and now and then extremely savage, and I saw a man get his shoulder dislocated after a violent fall. Still I sat watching it with great delight, sometimes for hours, as certainly they played it with their whole souls, shouting and yelling, leaping like

* Es Shark, the East. It is possible that the word Asia was derived from it when it is remembered that the Greeks and Romans must have had great difficulty in pronouncing both the Arabic gutturals and semi-gutturals.

roe, and everybody playing off side when it seemed good to him, and glorying in his crime.

Towards midday came the Chamberlain, bringing back our guns with many thanks and offers of purchase, which we had to decline, as neither of the guns belonged to us. With him he brought a double-barrelled hammerless gun in good condition, and with the maker's name (Green, Haymarket, London) engraved upon it. He said it was the Kaid's, who set great store by it, having received it as a present from a merchant on the coast, and specially he wished to know if the gun was what would be called of first-rate workmanship in England. I told him that it was and probably cost about twenty pounds, and that the son of our Sultana could buy no better or more expensive weapon, unless, which I said did not seem probable to me, he had his guns adorned with gold or precious stones.

But better than the guns, or talk of guns, was the invitation which he brought from the Kaid, saying he would like to see us in the afternoon. As such an invitation was, in our position, really a command, I hesitated some time before accepting it. The Chamberlain saw what was passing in my mind, and to gild the pill, remarked that had the Governor not been suffering from his wound, he would have got upon his horse and ridden to our tent. Though I felt sure that what he said was quite untrue, still mankind is so constituted that humbug flatters us, even although we think we see through it. So I accepted, and Si Mohammed departed after many compliments and with a promise to come and fetch us, and usher us into the Presence, in the afternoon. Lutaif lamented bitterly that we had no European clothes with which to endue ourselves, and properly impress the Kaid.

It must not be forgotten that in the East (and Mogreb-el-Acksa, though it means Far West, is perhaps as Eastern as any country in the world) European clothes, hard hats, elastic-sided boots, grey flannel shirts, with braces, mother-of-pearl studs, two-carat watch-chains, and all the beauty of our meanly contrived apparels, are to Mohammedans the outward visible sign of the inward spiritual Maxim gun, torpedo boat, and arms of precision, on which our civilization, power, might, dominion, and morality really repose. A shoddy-clad and cheating European pedlar, in his national dress, always suggests to Easterns the might of England somewhere in the offing, and though they laugh at the wearer of the grey shoddy rags behind his back, they yet respect him more than if he were attired in the most beautiful of their own time-hallowed garments, which they know no European puts on but for some purpose of his own. But if a European loses respect in wearing Moorish clothes, he gains in another way, for the Moors are constituted like other men, and, seeing a man dressed in the clothes they wear themselves, converse with him more freely, even if, as in my case, his knowledge of the language is so slight as to make conversation through an interpreter a necessity.

So we put on the best we had all cleanly washed, and

Lutaif arrayed himself in a brand new white Selham
(burnous), and looked more Biblical than ever as he stood
forth to be my Aaron, I having resolved, in order not to
make myself ridiculous, to refrain from saying anything
in Arabic, unless I saw a chance to get some phrase in
pertinently, and with effect.

Punctually at half-past two the Chamberlain, accompanied
by a single follower, came for us, and we—that is, Lutaif,
myself, Mohammed el Hosein, and Swani—walked as
majestically as we could across the deserted Maidan, baking
in the sun. We passed through several courts in which our
friends the horses and the mules were tied, and I observed
the wounded cream-coloured stallion of the Kaid tethered
alone and guarded by a little boy who flapped the flies
away with a green bough. Passing by the door of the
Mosque, we saw a preacher holding forth to a congregation all dressed in white and seated on the ground. No
coughing drowned his saw, no shuffling of chairs disturbed
his eloquence, the listeners sat as solid as limpets on a rock,
whilst his voice rose and fell in measured cadences, reminding one of the long rollers in a calm, just'off the line.
The door of the mosque was a poor specimen of the bronze-
plated work adorned with pious sentences, which can be
seen to such perfection in the mosque at Cordoba; the
knocker of the familiar round Arab pattern, which the
Moors have left in half the houses throughout Southern
Spain. A narrow passage, where a few Jews and tribesmen
sat waiting for an audience, led beneath a horse-shoe arch-

way. Then, climbing up a dark and almost perpendicular staircase, we emerged into a lofty ante-room where several men sat on the floor preparing saffron, which covered half the room with a dense carpet of bright purple blossom, whilst in a corner lay a clean white sheepskin with a mass of orange saffron fibres all gathered in a heap. At one end of the room there was a narrow doorway, where two men with long guns in their hands kept watch, and people going out and in continually; some emerging crestfallen, and others radiant, as in the times when kings, even in Europe, gave personal audience, and their subjects spoke with them face to face.

Here we waited almost half an hour, no doubt on purpose to impress us with the amount of business which the Kaid had to transact. For myself, I was not sorry, as I had full leisure to observe all that was going on. Though all the people in the room and the two guards must have known who we were, no one showed curiosity, and one man talked to me, pretending to comprehend all that I said as if he wished to put me at my ease. We slipped our shoes off at an intimation from our guide and entered the Presence Chamber, a narrow room with an " artesonado "* ceiling in the Hispano-Moorish style, but vilely daubed in Reckitt's blue and dingy red, and with cheap common gilding making it look tawdry and like the ceiling of an old-fashioned music-hall.

* Ceilings divided into an infinity of little squares, and with pendant knobs here and there, and much inlaid work and gilding, are called " artesonado " in Spain, and I know no word in English by which to render it.

In a recess within the wall two boys were sitting doing nothing in rather an aggressive way. To my eye they looked rather androgynous, but not more so than many young men one sees in Piccadilly on a fine afternoon; and who would tolerate even a suspicion about the noble Shillah race!* The room was carpeted with fine, almost white, matting, over which here and there were thrown black and white rugs from Sus, all worked in curious geometric patterns, woven from the softest of wool mingled with goats' hair, and with long fringes at the edge.

Upon a dark red saddle-cloth† and using an angle of the wall to lean against, his wounded leg stretched out before him on a sheepskin, and with cushions at his back, his Excellency sat. Luckily Arab manners (and in these matters Berbers follow the Arab lead) prescribe no Kiddush, or, most infallibly, situated as we were, we should have been obliged to make it, with the best grace we could. So we advanced, were formally presented by the Chamberlain, shook hands, and after being greeted quietly, but courteously, and after Lutaif had answered quite in the

* "Shillah race," see books on Morocco, written sometimes by those whom Disraeli described as " flat-nosed Franks," and who, no doubt humiliated by having met in the Arabs a finer type than their own, turned to the Berbers with the relief that the earthen tea-pot must find when taken away from the drawing-room companionship of " powder blue " china, and put back again on the kitchen dresser.

† These saddle-cloths, called in Morocco " libdah," are carried by respectable Moors when going to the Mosques to pray; Talebs and men of letters (who ride mules) generally have one loose upon their saddles, to sit upon when they dismount. Men of the sword disdain them and use them only underneath their saddles, where they place seven of them, of several colours, blue, red and yellow, and add an eighth, when on a journey, of white wool and separate from the others (which are all sewed together), so that it may be removed and washed.

style of Faredi, sat down upon a rug and leaned against the wall, tucking our feet well underneath our clothes to show our breeding, and remained silently waiting to hear what the Kaid had to say.

Mohammed el Hosein and Swani advanced, lifted the Kaid's selham, kissed it, and then retreating sat down, so to speak, below the salt, whilst in the doorway the two sentinels stood as unmoved as if they saw a Christian every day. Two or three elders sat round the room as stolid as josses in a temple, two Talebs, besides our friend the " Taleb of the Atlas," were writing letters, and the Chamberlain stood at attention till the Kaid waved him to take a seat.

No doubt his Excellency took mental notes of us, and certainly I looked him over carefully, thinking that in a personal discussion upon horseback, out on the Maidan, he would prove a very awkward foe.

Just about forty years of age, thick-set, and dark complexioned, close black beard trimmed to a double point, rather small eyes, like those of all his race, he gave no indication of the cruelty for which he was renowned; not noble in appearance as are many of the Sheikhs of Arab blood, but still looking as one accustomed to command; hands strong and muscular, voice rather harsh, but low, and trained in the best school of Arab manners, so as to be hardly audible. Just for a moment, and no more, I got a glimpse of the inside man as I caught his eye fixed on me, savage yet fish-like, but in an instant a sort of film seemed to pass over it, not that he dropped his gaze, but seemed deliberately to veil it, as if he had reserved it for a more fitting opportunity. By race and language he was a Berber, but speaking Arabic tolerably fluently, and adapting all his habits and dress to those in fashion amongst Arab Sheikhs. His clothes white and of the finest wool, and clean as is a sheet of paper before a writer marks it black with lies. The Talebs never stopped opening and writing letters, now and then handing one to the Kaid who glanced it over and said " Guaha " (" Good "), and gave it back to have the seal affixed with one of the three large silver seals which stood upon a little table about six inches high. The sealing-wax was European, and kept in a box of common cardboard, which had been mended in several places with little silver bands to keep the sides together, as we should mend a lacquered box from Persia or Japan. Behind the Kaid, to mark his seat, upon the wall were painted three " ajimeces,"* roughly designed in blue and red and green in the worst of taste. For furniture, in addition to the matting and the rugs and leather-covered cushions, the cover cut into intricate geometric patterns, the room contained a small trunk-shaped box (perhaps entirely stuffed with gold, Allah hualem), a Belgian single-barrelled nickel-plated breech-loading gun hung on a nail, and the before-named double-barrelled English gun (from the Haymarket of the mysterious Londres or Windres, in the Isle of Mists),

* An ajimez is one of the little long-shaped horse-shoe windows, so frequently seen in Moorish buildings; often in the sides of towers, as in those of the Giralda at Seville.

a large pair of double field-glasses; some bags of hide, two porous water bottles, a bundle of reed pens, and two or three pieces of bread, the staff of life, which fills so large a place in Moorish thoughts and life, and which an Arab of the old school breaks, but never touches with a knife. Two negro boys with dirty handkerchiefs, and boughs of walnut, stood on the right and left hand of the Kaid, and flapped away the flies.

Oh, what a falling off from when, in Medina el Azahra, the great palace outside Cordoba, the Greek Ambassador beheld the Caliph's court, the wonders of the great gold basin filled with a sea of quicksilver, and the slave boys, beautiful as angels, who fanned their lord with jewelled fans made of the feathers of the wondrous bird from Hind, which on its spread-out tail carries a hundred eyes. But in Kintafi, even the Kaid himself held in his hand a branch torn from a bush, and flapped occasionally with his own august hand, when the myriads of flies became impertinent.

People were going in and out perpetually, like bees into a hive, or politicians pretending they have important business in the House of Commons. Some brought petitions, others begged for mules, horses, a gun, or anything which came into their minds, and generally the Kaid gave something, for Moors all pique themselves upon their generosity.

Besides suppliants, Jews and various artificers were hanging about the ante-room. A silversmith advanced to show a half-completed silver-sheathed and hafted dagger, engraved with pious sentences, as " God is our sufficiency and our best bulwark here on earth," and running in and out between the texts a pattern of a rope with one of the strands left out, which pattern also ran round the cornice of the room we sat in, and round the door, as it runs round the doors in the Alhambra and the Alcazar, and in thousands of houses built by the Moors, and standing still, in Spain. The dagger and the sheath were handed to me for my inspection, and on my saying that they were beautifully worked, the Kaid said, "Keep them," but I declined, not having anything of equal value to give in return, and being almost certain if I sent a present from Mogador, that it would never reach its owner's hands. So we gravely put the dagger backwards and forwards with many courteous waves, " It is yours, take it I pray, although unworthy your acceptance "; and I " The dagger is in worthy hands, let it remain with one who had the good taste to order such fine work, and has the hands to use the weapon when there is need." A pretty little comedy, my share of which I conducted through Lutaif, not wishing to fall into barbarities of speech and make myself ridiculous before so many well-spoken men.

Slave boys, in clothes perhaps worth eighteenpence, served coffee, rather an unusual thing in visiting a Moor, for all drink tea. The tray was copper, beautifully chased, and adorned with sentences from the Koran, the service varied, and consisting of a common wine glass, one champagne glass of the old-fashioned narrow pattern, three cheap French cups, and a most beautifully engraved old Spanish glass goblet out of which his Excellency drank. The coffee-pot looked like a piece of Empire silver ware; the coffee excellent, and brought most probably by some pilgrim from Arabia, and used only on great occasions such as the present, or at a marriage feast.

The talk ran chiefly upon our journey: why had we come? why dressed like Moors? where were we going? and why we had no letter from the Sultan; and, above all, why had we not called at his house in passing as was usual for all Moors (of our assumed condition) to do when on the road? I answered that we were going to Tarudant, that we were dressed as Moors because the people were not accustomed to see Christians, and might have insulted us; and that we did not call upon him knowing he had so many visitors, and not wishing to intrude. As to a Sultan's letter, that was unnecessary, for I knew well if I had one he would find some good reason to stop us, under the pretext that the roads we should encounter would be unsafe. Moreover, that I had travelled much in Morocco, and did not like to have a Sultan's letter, for if I had one, no one would let me pay for food, and that I could not bear to be a burden on the poor tribesmen amongst whom I passed.

My object in visiting Tarudant seemed to him incomprehensible, as it was merely curiosity, and for a moment it crossed my mind, should I make up some reason, such as a vow to make a pilgrimage, a wish to see if there were mines in the vicinity, or something which should seem sufficient in his eyes? but in a minute was glad I had not done so, for he asked, did I know the English adventurers who, a few months ago, had tried to land upon the coast of Sus? As at that time I did not, I answered that they were personally unknown to me, but that I totally disapproved what they had done, especially because our Government had warned British subjects not to try to come to terms with the Sus chiefs, and that the Sultan had expressed a wish that nothing of that nature should be done. I added that personally I reverenced all Governments, especially my own, having been once a member of the great Council of our Empire, which, I took care to state, with all the patriotism I could command, was, on reliable authority, said to be the largest and finest in the world. He answered " Guaha, that is so. Allah himself appointed Governments, placed the sword of justice in their hands, and it is for them to say what should be done and see their wishes are respected." To this I gave assent, and he inquired was I still of the Council? and, when I answered no, asked if I had quarrelled with the Vizirs, or done anything unpleasing to them, or was I only tired of public life? Finding our parliamentary system too intricate to explain, I said I was tired of the cares of state, and he replied, " Yes, they are heavy, and I myself have never wished to go to Court." As I knew well if he ever ventured there his life was not worth a rotten egg, I applauded his resolve, spoke of the pleasures of a country life, and, as all hitherto had passed through the good offices of Lutaif, thought that my chance had come, and mustering up my Arabic told him he should be content with what God gave him, for as he was, he was a Sultan in himself. He smiled, whether at the compliment or my bad Arabic, I do not know, and beckoned to Mohammed el Hosein to come and speak with him.

Mohammed el Hosein advanced, kissed his selham, and in an instant became a gentleman and conversed on equal terms. What they conversed about I do not know, as all their talk passed in Shillah; but I conclude the Kaid was satisfied with what Mohammed said, for, signing to the slave who poured the coffe out (a knave who had a heavy silver earring in his ear, from which depended a cross-shaped ornament with Solomon's Seal engraved upon it), he told him to give Mohammed coffee; he did so, in a white egg-cup, which, as it stood behind the coffee pot, I had not previously observed. " Do you in Europe travel about all through the different countries without letters from your Queen?" " No," I rejoined, " we take a letter signed by our Grand Vizir, and show it, if asked for, at the frontiers of the various States." " Of course you have one?" he immediately replied. I answered " Yes," and just remembered I had left it behind me in the hotel in Mogador; but luckily he did not ask to see it, or I should have had to show him a letter which I had with a large seal upon it, which probably would have answered just as well.

I pressed him to allow us to go on to Tarudant; but he became mysterious, said the roads were bad, the people dangerous, and that to save our lives he had acted in the way that he had done.

Nothing is so disagreeable as to have your life saved in your own despite. Fancy the feelings of a would-be suicide when some intruding fellow, like a great Newfoundland dog, jumps in and pulls him out, and then on landing asks him for his thanks!

After the coffee, talk ran a good deal upon various things, polygamy and monogamy, always an interesting subject to all Orientals, who, being primitive in tastes and habits, set much account on primary passions (or affections) and think more of such matters than we do, talking quite openly and without periphrasis on things we do, but never talk about, or if we do, lower our breath in talking. Strange and incomprehensible to a logician that a man should say, I am hungry, thirsty, tired, and think there is something wrong, indelicate, or indifferent in mentioning the kindred passions, presumably implanted in his body by the same All-Wise Creator who endowed him with the capacity to feel thirst, hunger, or fatigue. The Kaid was of opinion that polygamy was natural to mankind, and asked me if the English did not really think so in their hearts. It is most difficult, without having been duly elected, to speak for a whole nation, so I replied that many acted as if they thought polygamy was right, but I ventured to opine that advanced thinkers in general inclined to polyandry, and that seemed to be the opinion which, in the future, would prevail. This he thought clearly wrong; but I explained that advanced thinkers were inclined to hold that women could do no wrong, and that all infamy of every nature had its root in man.

The prisoners in the Riff* next were enlarged upon, and the Kaid asked if they had been released, and what I thought about the whole affair. Thinking the opportunity favourable to air my Arabic again, I said laboriously that I had heard there were some prisoners in the Riff, and added that there were prisoners also in the Atlas, but no doubt the Sultan would soon order their release.

His Excellency's wounded leg was, on the whole, the subject which gave most scope for talk. Neither his Arabic nor mine was fluent enough to explain, or understand quite fully, what had taken place. More coffee having been ordered, the Chamberlain entered into an explanation which Lutaif translated when I (as happened now and then) became bewildered in the current of his speech. About two months ago the unlawfully begotten people in the Sus, egged on by certain British traders,† had rebelled against their Lord. The chief offenders were sons of Jews (Oulad el Jahud), who had withdrawn themselves into a fortified position three or four days' journey from Thelata-el-Jacoub.

---

* At that time, October, 1897, several Spanish, Greek, and Italian sailors were detained by the Riff tribes, having been captured when their vessels were becalmed near the shores of that province. European diplomacy having, as usual, failed, a Jew from Tangier with the aid of the French Consul in Tangier arranged for their liberation, and they arrived in Tangier on the same day that I arrived from the Atlas. The populace, chiefly Spaniards from Malaga, who had "had trouble" (knife thrusts given and received), welcomed us with acclamation, that is, they stared at us and shouted.

† The Globe Venture Syndicate, I i magine.

The Kaid was ordered to co-operate with the troops in Sus and bring the rebels back to allegiance, or destroy them all. Most probably the Kaid had no objection to an expedition out of his territory, though in point of fact his own allegiance to the Sultan did not much trouble him. However, mounted on the white horse, which I saw wounded and drinking in the river, and leading on his men, the Kaid had advanced against the revolted tribesmen, who were strongly posted amongst rocks protected with an outwork of Zaribas* made of prickly bushes, from behind which they fired upon the Kaid's forces, who had no shelter, and soon suffered heavy loss. The saddles emptying on every side, the Kaid was left almost alone with about twenty men, amongst whom was I who speak (the Chamberlain remarked in passing, but without any self-consciousness), his horse received a bullet in the chest, and another in the head; but still the Kaid advanced, keeping his horse's head as much as possible between him and the fire. At last another bullet struck the horse close to the nose, and he wheeling, the Kaid received a bullet in the left leg and fell! "Then," said the Chamberlain, "I rode close up to him, and the bullets tore up the grass on every side, when our men rallying brought us off, I and four others carrying the Kaid under a heavy fire, and the white wounded horse walking beside us, till we reached our camp."

During the tale the Kaid sat imperturbable as a joss cut out of soapstone, but punctuating all his henchman said

* Corrals, or enclosures.

with an occasional " Guaha," or some pious ejaculation fit
for a man of quality to use.

In the camp they placed the wounded Kaid upon a mule,
and fighting for the first two days almost incessantly,
upon the evening of the sixth day they brought him home,
two slaves having supported him on either side stretched
on the mule, too weak to sit upright, and with four more
helping the wounded horse, which the Kaid on no account
would leave to be the prize of Kaffirs such as those who
dwelt in Sus.

My opinion of all concerned rose not a little on listening
to the history and on learning that the Kaid had hesitated
not an instant to sacrifice his life, and those of all his followers, to save his favourite horse. And all the time the tale
was going on I thought where had I heard all this before,
for every incident seemed to me in some strange way
familiar. At last I recollected that Garcilasso de la Vega
(Inca) in his " Comentarios Reales del Peru," when he
relates the civil wars between the followers of the Pizarros
and the forces of the Viceroy, tells how Gonzalo de Silvestre,
after the battle of Huarina, found himself alone upon a
horse wounded twice in the head, and in the chest, and that
he gave himself up for lost, thinking his horse would fall,
when " feeling him a little with the bridle, the horse threw
up his head, and, snorting, blew blood through his nostrils
and seemed relieved, then went on galloping, and presently
I passed one of our partizans retreating, badly wounded, on
a mule, not able to sit upright for his hurt, and by him

walked an Indian woman, with her hand upon the wound to stop the blood."* Gonzalo de Silvestre and the wounded man and horse all got off with their lives, no doubt the same tripartite deity assisting them who in his indivisible aspect came to the assistance of the Chamberlain and of the Kaid.

Could I then undertake to examine the leg and perhaps extract the ball ? was put to me through the medium of the Chamberlain. For a moment I hesitated, thinking that if the ball was near the skin I would hazard it, and so earn the eternal gratitude of the Kaid and be sent on to Tarudant with honour and with an escort of his followers to guard me on the way. One look dispelled my hopes, for the wound was high up in the thigh, close to the femoral artery and had almost healed, although the patient said it gave him pain, and stopped him from getting on his horse, though when once mounted he could make a shift to ride. Reluctantly I had to say I was not able to undertake so serious a case. The Kaid's face fell, so I advised him to send for an English doctor who I knew was staying in Morocco city for his health, and who would have been glad to see so strange a place, and put the patient upon his legs again. If he has done so by this time I do not know, or even if the Kaid made up his mind to send for him; but the chance for a doctor was unique, and therefore has, most probably, been missed.

Governors of provinces in Morocco, and throughout the

* Garcilasso de la Vega, "Comentarios Reales del Peru," Fifth Book, Part II, Chap. 21.

East, are rather shy of going to the capital, even in such a case as this; for once there, the Sultan often takes the opportunity of making them disgorge some of the money which they have plundered in their government. On the first rumour that the Governor is in disgrace, the tribe rebels, blockades the castle, burns it down if possible, and some neighbouring Sheikh sends to the Sultan and offers a large sum to be made Governor in the disgraced man's place. Even if things do not go quite so far as that, a journey in Morocco has its inconveniences, for generally the wives take the chance of the husband's absence to dig up his money and send it to their friends. This happened to the Kaid of Kintafi when wounded in the Sus, and he, on his return finding his money gone, divorced two of his wives, and treated all the others to some discipline, which the Chamberlain assured me had restored peace and order to his Excellency's house.

Our interview having lasted almost two hours, we rose to take our leave. I thanked the Kaid for his continued hospitality, assured him that I should not easily forget Kintafi, promised to send him a doctor if he wished, and quite forgot I was a prisoner. He on his part transmitted his good wishes through the medium of his henchman and Lutaif, and said he hoped we would not leave Kintafi for a few days more, as he was anxious to speak to us again. This was not quite the ending of the interview I had expected, for it amounted to an order we should not leave the place, so in conveying to him my best thanks for all his hospitality, I told him that I would let him know the latest news about

the prisoners in the Riff on my return to Mogador, and in
the meantime hoped Allah would guide him in all he did,
and that he would continue to dispense his hospitality to all
who passed, because, as Sidna Mohammed himself has said,
"that hospitality, even when unasked for, blesses both the
host and guest."

Lutaif, who had the pleasure of translating this farewell,
did not much like his task, but faced it manfully, and so
amidst a shower of compliments we took our leave, and left
the presence of our illustrious host for good. Still an experience not to have been missed, and differing extremely from
the ordinary visit paid by the travelling European to a
Moorish Kaid on equal terms, that is, when dressed in
European clothes, furnished with letters from the Sultan
and the ambassador of the traveller's country, when one
drinks tea, exchanges compliments, and learns as little of the
real go on of an Oriental house, as does a man born rich
learn the real workings of the people's minds with whom he
lives his life.

An Eastern potentate of the Arabian Nights was the
Kaid, with all the culture of the Arabs of the Middle Ages
absent, but as he was, the arbiter of life and death in a wide
district. A gentleman in manners, courteous to those whom
he had all the power to treat with rudeness or severity; a
horseman and a fighter; a tyrant naturally, as any man
would be if placed in his position; but no more tyrannical
in disposition than is some new elected County Councillor,
mad to make all men chaste and sober by some bye-law or

another; himself a victim to a lewd Puritanism, and an insatiable love of cant. Half independent of the Sultan, leading his own troops, dispensing justice, as he thought he saw it, in his own courtyard, and to me interesting in special as a sort of after-type of those great Arab Emirs who sprang from the sands of Africa and of Arabia, shook Europe, flourished in Spain, built the Alhambra and Alcazar, gave us the Arab horse and the curb bit, and kept alive the remains of Greek philosophy in Cordoba and in Toledo, when all the rest of Europe grovelled in darkness; then by degrees fell into decadence, and sank again into the sands of Africa, to still keep alive the patriarchal system, the oldest and perhaps the best conception of a simple life mankind has yet found out. Allah Ackbar; lost in a wilderness of broadcloth, I still praise God that such a man exists, if only to contrast him in my mind with the self-advertising anthropoids who make one fancy, if the Darwinian theory still holds good, that the God after whose image the first man was made had surely been an ape. Through passages and courtyards we reached the open space on which our tent was pitched, escorted by a guard of men well armed with guns and daggers, which appendages made them none the less loth to take a tip on coming to the tent than if they had been so many gamekeepers, who take their unearned money after a grouse drive, or a hot corner in the coverts, with an air of doing you a service, and whose contempt for you is only equalled by your disgust both at yourself and them.

But, over everyone a change had come, for we had stood

before the face of the great man with honour, and those who scarcely in the morning returned our salutations, gravely saluted us and condescended to enquire after our welfare and our health.

Swani and Mohammed-el-Hosein were radiant, more especially because the Kaid had sent a sheep, which they had already slain and given to a "master" (Maalem) to roast *en barbecue*. Although I personally was disappointed that we had not been able either to get an answer from the Kaid as to our return, still less to get permission to go on, yet I was glad to have seen him, placed as I was, and wondered if an English Duke in the Georgian times would have treated an Arab wandering in England, and giving out he was an English clergyman, as well as the wild, semi-independent Berber Shiekh treated the wandering Englishman who assumed to pass, not merely as a clergyman, but as a saint.

Four men appeared bearing the sheep on a huge wooden dish, smoking and peppered so as to start us sneezing; and when the Maalem had torn it into convenient portions with his hands, we all fell to, Lutaif and I with an appetite that civilization gives for such a meal; the rest like wolves, or men remembering the Hispano-Moorish proverb to the effect that meat and appetite go not always together, though both are sent by God.

Chapter X

EARLY next day the effects of our audience began
to manifest themselves. The sick, the halt, the lame,
and impotent, all besieged the tent, having been
kept away apparently by the uncertainty of the Kaid's attitude towards the Nazarenes. But as all Europeans are supposed not only to know something of medicine but to carry
drugs about with them on all occasions, the afflicted fairly
besieged me, and I dispensed my medicines with a freedom
quite without fear of consequences, and not restrained as
doctors are in " policed " countries, where every now and
then the public fall into a panic when a case of human vivi-
section, carried out (upon the poor) in the pure joy of scientific life, leaks from inside the precincts of some hospital.
Gratitude, which dies with knowledge, but flourishes luxuri-
antly as long as the medicine man is a being quite apart,
working his wonders unrestrained by scientific bounds,
showed itself in several ways. Yet, as per usual, most patheti-
cally in inverse proportion to the riches of the patient, for
several who should have recompensed my skill (or zeal)
according to the goodness of their clothes, slunk off, as
people do at home after street acrobats have been performing, and when the little boy who has risked his life upon a

pyramid of father, big brother, and several uncles, straining in his baggy cotton tights (which give him the appearance of a cab horse shaken in the legs), is just about to come round holding out his tambourine. So I dispensed my stock, which I had brought to spread my fame and smooth my path in Tarudant, quite cheerfully; ophthalmia, tetanus, sciatica, elephantiasis, ulcers, and twisted limbs, with rheumatism, deafness, and El Burd, I alleviated by the faith of those who took my drugs. The more extraordinary and complicated were the instructions which I gave, the better pleased the patient was, and sometimes came back twice or three times to ask if the quinine or Seidlitz powder was to be taken on every fifth or ninth recurring day. One poor Bezonian, for whom I had prescribed something or other, came in the evening, and lugging out from underneath his cloak a dirty pocket-handkerchief, produced a handful of greasy copper coin, worth, perhaps, one penny halfpenny in all, and with excuses for his poverty in Shillah, which came to me through Arabic, entreated me to pay myself for my prescription, I answered as nobly as I could, " The credit is Allah's," and the poor man advanced, kissed me upon the shoulder, and went out, perhaps to tell today of the great Christian doctor who would take no fee. Others brought eggs and bread, and these I took, as it would have been an insult to them to refuse, and besides that, I fancy the mere idea of having paid made my hell broths appear more efficacious to the simple folk. Who, even in England, that does not believe that the two guineas which he pays to see some

great specialist * in Harley Street does not advance his case,
and were it five I fancy the greater part of patients would
leave the doctor's mansion cured, or else omit to go, and all
is one in cases where faith heals.

My most popular recipe for coughs may yet achieve the
popularity which is reserved for faith-healing amongst the
Christian world. Take four Beecham's Pills, and bruise them
in a mortar with an ounce of cloves and two of Argan oil,
a piece of rancid butter, and a cup of magia (spirit made
from dates); rub well upon the chest, anoint the feet, and
take a spoonful of the same liquid in tepid water, now and
then, continuing our light and nourishing couscousou and
shisha,† and please Allah the cough will disappear.

Tired of dispensing, I strolled out to the olive grove, sat
down, began to smoke, and watched two men seated close
by, dressed in white robes, and evidently of the richer sort.
One read a letter to his friend, with explanations upon every
line, and with apparently some trouble to himself, for every
now and then he drew a character in the sand with his
forefinger, and compared it with the doubtful character in
the document he read. In the same fashion I have seen grave,
reserved, grey-bearded men in South America sit entertained for hours in " painting "‡ horses' marks upon the

---

* A philosopher has remarked that liar, damned liar, and skilled witness stand
in a progressive ratio, and for all I know, " specialists " may hold much the same
position in the world of medicine.

† Shisha is a kind of thin porridge not unlike the skilly of our Christian
prisons.

‡ " Painting a mark from the Spanish expression 'pintar una marca.' In the
same way in Western Texas ' Pinto,' a piebald horse, became ' paint,' ' cabresto,'
a halter, ' cabress,' and so on, in all conscience and tender heart."

sand, and reasoning wisely upon every one of them. The
man who listened said not a word, but looked entranced
with admiration at the deep knowledge of his friend. I take
it, reading and writing should not be abused, or it may
chance with them to fare even as it has fared with sweet
religion, which first a mystery concealed in a learned tongue,
and thus respected and believed without inquiry, then became understandable by its translation to the vulgar speech,
lost credit, and today has fallen into a fashion, and changes
in complexion, form, and authenticity, as quickly as
a hat brought from Paris in the spring falls into dowdi-
ness and becomes ridiculous almost before the owner is
aware.

   We marvelled greatly that we had not been able to discover what were the Kaid's intentions in regard to us, and,
casting up the time, found we had passed nine days already
in the place. Curious, in prison, or in a ship, or stuck alone
in some wild hut on prairie or on Pampa, to remark how
long the time seems for the first few days, and then begins
to race, so that before one is aware a month is past which at
the first looked like eternity to face. And stranger still, how
in a week or so, the newspapers and books, the so-called
intellectual conversation, news of the outside world, the
theatres, churches, politics, and the things which by their
aggregated littleness, taken together, seem important, fall
out of one's life. The condition of one's horse, the weather,
crops, the storm, or coming revolution, all take their places
and become as important as were the unimportant great

events which a short time ago, served up distorted in an evening paper, whiled away our time.

Thus life at Kintafi after a little became quite natural to us, and at least as cheerful for a continuance as life in Parliament, in Paris, London, or any other of the dreary hives of pleasure or of thought. We rose at daylight, drank green tea and smoked, went down to bathe, came back and breakfasted, looked at the horses led to water, listened to the muezzin call to prayers, walked in the olive grove or watched the negroes in the corn field; engaged in conversation with some of the strange types, we read el Faredi, speculated on how long the "rekass" would tarry on the road from the Sultan's camp, and wondered at the perpetual procession of people always arriving at the castle to beg for something, a horse, a mule, a gun, some money, or in some way or other to participate in the Kaid's Baraka.* Had I but been allowed to ride about and explore the country, I should have been content to stay a month. However, there was no order, and all those who are not strong enough to disobey have to stick strictly to an order in the East.

As an example of how orders are obeyed, one day during my sojourn at Kintafi, Lutaif and I had wandered about a mile following the Wad el N'fiss and crossing it once or twice to save the bends. As we were walking we had to take

---

* Baraka literally means a blessing. It is also used in thanks, as *Baraka Lowfik* ("The blessing of God be on you"). And not infrequently as a sort of general term for goodness or generosity. Arabs rarely say "Baraka Lowfik" in thanking a Christian, but use the less religious phrase "Kettir heirac." Neither do they (in Morocco) ever salute a Christian with *Salam Aleikoum* ("Peace be with you"), as peace is only for Believers.

our slippers off and cross barefooted, picking our way over the pebbles in the fierce stream, which made it difficult to walk. On our return, just at a ford where the current ran particularly swift, we met two mounted men, followers of the Kaid, whom we knew well, and one of whom I had prescribed for in my character of medicine man. Thinking the chance a good one, I asked the man to let me ride across, knowing his feet were hard as leather, and intending to have given him a trifle for lending me the horse. The man excused himself with many apologies, and said he had been sent to exercise the horse, and had no order to let any one ride on it, and dare not upon any pretext let me get upon its back. He had no reason to be uncivil, and was no doubt in terror of what the Kaid might do had the news come to him that he had gone an atom beyond his strict command.

Easterns of any nation are good company to be thrown with on an occasion such as the one in which I found myself. Lutaif had a never failing fund of stories about things he had seen and heard, and told them with the absolute lack of self-consciousness which alone makes a story pleasant, and which distinguishes an Eastern story-teller from the Western, who in his story always has an eye on the effect of what he tells.

At present in the Lebanon it seems there is an exodus of all the educated young men towards America; and in New York there is a Syrian quarter where they speak Arabic, carry on small industries, and, curiously enough, are known to the natives as " the Arabians," a designation which must sound most strangely to a Syrian's ear. One of these Syrian young men, but in this instance quite uneducated, and speaking only a rough Arabic *patois*, started to try and reach America, where the streets are paved with gold. Having shipped as a deck hand on board a steamer at Beirout, he reached New York. There he was put ashore, and failed to get employment; wandered about the streets; was taken to the Turkish Consul, and by him shipped to Marseilles as a vagrant who was unable to support himself in the free land where every man is better than his neighbour if he has more money in his purse. Dropped ashore at Marseilles, he got aboard an Italian steamer going to the River Plate, was found and flogged, then fastened to the mast for six or seven hours, and when the vessel touched at Malaga, shoved ashore, after receiving several hearty kicks. There, almost starving, and ignorant of Spanish, he set out to tramp to Gibraltar, about sixty miles away. But most unluckily he reckoned without the *odium theologicum;* for happening to wear a Turkish fez, and Spain just at that time being engaged in a squabble at Alhucemas, in the Riff, the country people were in a fervour of religious rage against the Moors. New York, Marseilles, the Italian steamer, all were as nothing to what he had to suffer in his tramp in the land specially the Blessed Virgin's* own. No one would have him in the villages, and when he asked for bread he met, in the fashion of the Holy Scriptures, with a stone. The children

* Andalusia is known to Spaniards as La tierra de Maria Santissima.

hooted him, the shepherds minding their sheep slung pebbles at him, and after three days' journey, in a most miserable plight, he reached Gibraltar, where he met a Moor, who gave him a bag to carry to a certain place. The bag turned out to contain cigars which the Moor wished to smuggle on board a ship. Up came a policeman, beat him a little with a short club (as he explained), and took him off before a magistrate. There not a soul spoke Arabic, so he was remanded to the cells, where, as he said, he was quite comfortable and better off than in that cursed "town called Spain." At last an interpreter arrived and the poor man found himself free, but starving, and in despair crossed to Tangier, and there on my return I found him saving up coppers to pay his passage once more to New York.

Back in our tent the staring recommenced, as for some reason many of the wilder tribesmen were, so to speak, in town today. They sat outside the tent like sparrows on a telegraph, and looked at us as if we were the strangest sight they ever had beheld. Even the stolid glare of a hostile (or stupid) audience at a public meeting was nothing to their gaze. At times I thought had we but brought a monkey and an organ our fortunes had been made, and we should have to buy a camel to carry off the avalanche of copper coin.

Still there was no sign of the rekass; and so the 28th and 29th slipped past, leaving us still a-thinking, still cavilling, and wondering how much longer we should have to stay.

Almost the most interesting, and certainly the most

pathetic, of my patients arrived during the 29th. A long, thin, famine-stricken man dressed in rags begged me for medicine for a " sad heart," and certainly he had good cause for sadness, though I fancy that the peseta* which I gave him may have done him at least as much good as the last of my quinine. It appeared that the late Sultan, Mulai el Hassan, had destroyed his house, taken his property, and driven him to exile. Quite naturally the present Sultan had too much filial respect for his late father to undo any action, just or unjust, that he had thought fit to do. Therefore my patient, whose " sad heart " had stood out for three whole years, was still a suppliant, and his present errand was to try and interest El Kintafi in his case. For six long months he had been in the place trying his luck without success. Sometimes the Kaid would promise him his help, and then again tell him to come when he had thought the matter over and resolved what was the best to do. Meantime the man slept in the mosque by night, by day stood at the gate, and when the Kaid rode out clung to his stirrup and implored his aid. He said, " I see him every six or seven days, but there is no hope but in God." Still he was cheerful, had his rags well washed, and was as resigned and dignified as I am certain that no Christian, out of fiction, could possibly have been. " God the great Helper "; but then how slow but merciful in this case, if only by the faith he had implanted to endure his own neglect. So the sad-hearted man of sorrows made

---

* The recent war having reminded the public that a country called Spain exists, I feel that I am not obliged to explain how little English money a peseta is worth at the present rate of exchange.

his notch upon my life, as the old Persian and the Oudad had done, and still perhaps waits for the Kaid on mornings when his Excellency rides out to hunt or hawk with a long train of followers, issuing from the horse-shoe arch, with negroes holding greyhounds in the leash, horsemen perched high on their red saddles, the sun falling upon long silver-mounted guns, haiks waving in the air, whilst from the ramparts of the castle comes the shrill note of joy the women raise when, in Morocco, men go out to hunt, to war, to play the powder;* or when, at weddings, the bride, stuffed in a gilded cage upon a mule, is taken home.

An aged Israelite with a long train of mules came from the Sus that morning. He wore a sort of compromise be-tween Oriental and European clothes, which gave him an incredibly abject look, the elastic-sided boots and ivory-handled cane contrasting most ill-favouredly with his long gaberdine; his ten-carat watch-chain, with a malachite locket hanging from it, rendering the effect of his maroon cloth caftan mean and civilized. He told me that his chief business was to lend money to the Kaids, and that his mules were packed with silver dollars, being the interest on the capital lent to various Governors in Sus. He expressed no fear of any attack upon his caravan; and when I quoted the

* The Powder Play (Lab-el-Barod) is known in Algeria as the fantasia. Both in that country and in Morocco it is the imitation of an Arab tribal battle. The horsemen rush forward and fire their guns in parties or singly, stand up on the saddle, fire under their horses' necks, and over their tails, throw their guns in the air and catch them, and perform all the evolutions which their ancestors per-formed with javelin and spear. This exercise prevailed in Spain till the middle of the last century under the name of the " Juego de Cañas." It is still played in the East with reeds.

saying that "if the caravan is attacked the poor man has nothing to fear," returned, " Nor has the Jew, who is indispensable to the great ones of the earth." Nevertheless he bore about him several old bullet wounds, and carried underneath his gaberdine a first-rate Smith and Wesson pistol, which he said he would not care to be obliged to use. I put him down as one who, given the opportunity, would shoot an unbeliever, like a dog, having generally observed that readiness to shoot goes in an inverse ratio with readiness to talk, and that the man who always has a pistol in his hand might just as often, for all purpose of defence, carry a meerschaum pipe. The way he travelled was curious, for, in the Sus amongst the Berber tribes, he had to take a tribesman, to whom he paid a certain sum, to see him safely through the tribe, who, in his turn, delivered him, on leaving his territory, to another man, and so on right through the country which he had to pass. This system is recognized throughout the Atlas range, and generally wherever the Berber tribes inhabit, and is known as el Mzareg, that is, the protection of the lance, for anciently the protecting tribesman bore a lance, but, nowadays, usually is satisfied with the stout cudgel which all hillmen use. Baruch, the Hebrew with the ivory crutch-handled stick, informed me that, in the Sus, nearly every Jewish family was obliged to have his corresponding Mzareg, who protected and also fleeced them; and that, in consequence, most of those " Pedlars of the ghetto," in spite of all their industry, were poor. A curious down-trodden race the Atlas Jews, ostensibly the slaves of

every one, and in reality their masters; for owing to the
incapacity for commerce in the Berbers, every affair where
money changes hands has to be brought about by the assistance of some quick-witted Jew. So, just as in Europe, though
without being in other respects superior to the races amongst
whom they live, the Atlas Jews control the warlike Berbers
as easily and as completely as their brethren control all
those with whom they come in contact on a business footing
throughout the world. Baruch had his home in Mequinez,
and was not from Toledo, but an Oriental Jew, his people,
as he said, having come into Morocco after the great dis-
persion, and he himself being of the tribe of Benjamin;
though, when I asked him how he knew, he said it was a
tradition in his family, and that the ancients never spoke
untruth. Into this matter I forbore to enter, and generally
gave an assent, quoting the Toledan-Jewish proverb, that
" if Moses died, Adonai still survived,"* which he at once
knew in its Arabic form, and asked me, as Israelites in the
East will often do if you appear to know a little of their
lore, if I, too, was of the chosen race. This worthy Baruch,
in appearance like a head cut out of walnut wood, set round
with fleecy wool, asked me, when passing Mequinez, to
remember I had my house there, and said that I should find
him, Baruch ben Baruch, as a father and a friend. Unfortunately, since then I have not passed Mequinez; but, if
I do so, hope to eat my " adafina " in my father's house.

These promises and resolutions one makes, in passing

* " Si Moshé murio, Adonai quedô."

through the world, to return to some place which has struck our fancy, to see some friend who lives ten thousand miles away, are like the apple blossoms blown across a lawn by a May wind; for are not dead flower leaves and broken promises but the illusions of a possibility which might have turned out bitter in the fruit ? So, our medicine done, our stock of patience almost exhausted, another day went past, leaving us no other resource to pass the time but to lounge up and down the Maidan, and when evening came welcome the sunset on the spacious amphitheatre of hills.

As the sun sank, the ochre-coloured earth began to glow, each stunted hill bush stood out and became magnified, the rose and purple streaks of light shifted and ran into each other; then faded into violet and pale salmon-colour haze and falling on the snow-capped hills lighted them up, making them reverberate the light upon the rose-red walls and yellow towers, so that the castle seemed to burn, and the muezzin upon his tower appeared to call the faithful to their prayers from a red stalk of flame.

Chapter XI

EARLY upon the morning of the 30th we were astir, and heard a report that a rekass had been seen entering the castle gates the night before. Still, everything went on as usual at Thelata-el-Jacoub, men came and went, tall Arabs and squat Shillah; our animals all stood dejected and half-starved; a little pup having made friends with my Amsmizi horse, who played with him in a perfunctory way. The prisoners on the flat roof sprawled in the sun, passing the Peace of God, on terms of absolute equality with other men, who paused and gravely gave them back their salutation; birds drank and bathed in the little mill-stream under the oleander bushes; butterflies, marbled and black-veined whites, argus, fritillaries, and others quite unknown to me floated along in the still air, or hung suspended over the petals of a flower, and the brown earth of the Maidan gave back the heat like a reflector.

Still no tidings of our long looked for orders to be gone. All our companions in adversity, the Persian, the man with the " sad heart," the tribesmen in the tent which had been pitched close to our own; even the three Sheikhs from Sus at last had gone, having, after their long delay, settled their

business and assured me of their friendship should I visit
Sus, their minds made up never again to venture inside the
lion's den. Our tents alone, like mushrooms, dotted the
Maidan, and over all the valley the late autumn sun shed
a deep calm, making the hills stand out as if cut out of cardboard in the middle distance of a theatre.

The enormous mud-built castle, with its flat roofs and
flat-topped flanking towers, took a prehistoric air, and for
the first time I imagined I was looking at some old Carthaginian building; for its architecture resembled nothing
I had ever seen, being as unlike an Arab Kaskah as to a
feudal castle in the north, with somehow a suggestion of a
teocalli, such as Bernal Diaz, or Cortes, described in Mexico;
or as if related in some curious way to the great buildings in
Palenque, temples at Apam, or the more ancient dwellings
of the Aztec race upon the Rio Gila. Sheep walked behind
their shepherds, who played upon reed pipes as in the times
of Hesiod; goats went to pasturage, the kids skipping and
playing with the ragged boys, whilst from the distance came
the wild sound of the Moorish Ghaita, which, like the bag-
pipes, is ever most alluring heard from afar; from the hills
floated the scent of thyme, germander, gum cistus, and the
aromatic undergrowth of this last outpost of the European
flora, jutting into the sands of Africa. The river prattled on
its stony bed, and on its bank the horses of the Kaid went
down to drink, and listlessly I strolled to look at them,
thinking, perhaps, that I should have to wait until the
autumn rains had made the mountain paths impassable, and

that the Kaid would be obliged to send me back against his will through the mysterious passes of the south.

As I mused, drinking in the strange semi-Arcadian, semi-feudalistic scene, the Chamberlain, attended by his guards, crossed the Maidan, came to my tent, and, sitting down, informed me that it was the pleasure of the Kaid that we should start. I asked had the rekass arrived from the Sultan; but the Chamberlain would give no definite reply, until I told him that our men had seen the messenger arrive. What were the exact terms of the Sultan's message I did not learn, but they could not have been extremely pleasant, either for me or for the Kaid; for in Morocco the Sultan never likes to be put in the position of risking complications with Europeans who might appeal to their home Government, and usually casts all the onus both of action and of failure on his Kaids.

Now was the time to make a last despairing effort to go on to Tarudant; therefore, I tried the Chamberlain in every way I could think of, but without success. Though I had not by this time half the money with me, I offered him a hundred dollars, to which he answered as before, "What is the use to me of a hundred or a thousand dollars without my head?" I shifted ground, and said I could not think of leaving without a personal interview with the Kaid to thank him for his hospitality. But here again the faithful Chamberlain was ready with a message, "that it pained his Excellency not to receive me, but his wound had broken out afresh, and that he sent many salaams and wished me a prosperous journey back to the coast." So as a last resource I asked the Chamberlain "What if I get on horseback, and ride straight on to Sus?" only to be met with a grave question if I thought I had the best horse in the valley, and if, supposing any tribesman was to fire by accident, I thought my clothes were stout enough to turn a ball.

Seeing there was no chance, I made a virtue of necessity, ordered my tent to be taken down, and all got ready for the road.

The Chamberlain behaved with infinite tact, allowing no one to come near us, either to steal or ask for money, and took his " gratification " with an air of having earned it by laborious work; then shook my hand, and said he hoped some day to meet me, and that I was to think he had only acted as he had by the orders of his chief. He then shook hands, first with Lutaif, then with Mohammed el Hosein and Swani, and took his leave, leaving us all alone on the Maidan, in a broiling sun, with nothing to eat, and ten or twelve hours of the hardest mountain roads to pass before it was possible to procure food for our animals or for our-selves. How all the others felt I do not know; but for myself, when I reflected on my journey lost, my twelve days of detention, and how near I had been to reaching Tarudant, being stopped but by the merest chance, I felt inclined to laugh. Knowing that almost all the houses were in ruins on the way, and that the only place where it was possible to buy barley for the mules was at a little castle called Taguaydirt-el-Bur, which belonged to the Kaid, and the Sheikh of which was known as a fanatical and disagreeable man, I sent Lutaif to the castle for the last time to get a letter from the Kaid to his lieutenant on the road. Quite evidently they were determined we should go, for in about an hour Lutaif returned with the letter and a negress carrying some bread and couscousou, on which we made a hearty breakfast, knowing that we should get nothing more till night.

When one has little to pack, arrangements for a journey are soon despatched, and we had nothing but our tent and a few rugs; no food, no barley for the mules, and the precious stock of medicines, which was to have made my name in Tarudant, was long run out. The men all worked like schoolboys packing to go home, especially Mohammed el Hosein, who had never thought to leave the place alive, or at the least to lose his mules, and be well beaten for his pains. His spirits rose enormously, and he assured me he was ready even now to try the road by Agadir. Swani sang Spanish and English sailors' songs, not to be quoted until the " woman movement " either makes women accustomed to the tone of much of sailors' conversation, or else refines our mariners and makes their talk more fit for ears polite, or hypocritical. Ali was like a man in a dream, and felt his mule all over to see if it had suffered by the long exposure to heat and rain, with barely any food. It winced, and kicked at him to show its love, and to assure him that its spirit was not injured by its fast.

So all being ready, and not a soul about, I mounted, took the black Amsmizi horse by the head and felt his mouth, touched him with the spur, and let him run across the empty Maidan, turned him and made him rear, and plunging down the steep path to the beach, just met the horses of the Kaid for the last time, being led out to drink. The wounded cream-colour, now almost quite recovered, stood up and gave a long defiant neigh, and as we rode under the castle walls upon the stony bed of the N'fiss, the figure of the Chamberlain appeared, and waved to us in a friendly fashion, so that my last impression of the place was his grave figure, silent and robed in white, and the fierce stallion neighing on the beach.

About a mile, following the river's bed, the trail leads through a scrub of oleanders, rises and enters a fantastic path worn in the limestone rock, cut here and there into pyramids and pinnacles by time, by traffic, and by winter rains, and looking something like the "seracs" formed in the ice upon the edge of certain glaciers. We followed it about a quarter of a mile, and turning, saw the castle of the Thelata-el-Jacoub for the last time.

But as I checked my horse, who, now his little spurt of spirit over, felt the twelve days' lack of food, and hung his head, I gazed upon the monstrous mud-built, yellowish-red pile, marked once again the olive grove upon the edge of the Maidan, just caught the mosque tower with its green metallic tiles, the cornfields, and the wild, narrow valley stretching to the snow-capped hills, the river like a steel wire winding in and out between its steep high banks, and in my heart thanked fortune which, no doubt for some wise (though hidden) purpose of its own, had kept me prisoner for twelve well-filled days in such a place. Lutaif, in spite of all his piety, as he took his last look at the valley of Kintafi, I fancy muttered something which sounded like an imprecation on Mohammed and his faith, but yet confessed that even in the Lebanon there was no valley wilder or more beautiful than that of the N'fiss. Mohammed el Hosein felt at his beard as if to assure himself it still grew on his chin, and without stint cursed Kaid and castle, tribe, place, and all the dwellers in it to the fourth and fifth generation of the sons of mothers who never yet said No. Swani was of opinion that of all men he had seen the Shillah were the most like Djins, and beat even the Jaui,* who, as all men knew, are sprung from monkeys, in ill-favouredness of face and wickedness of heart.

Our feelings thus relieved, we set ourselves to drive our half-starved beasts over the mountain roads. The miles seemed mortal, and the scenery, though wonderful enough, chiefly remarkable for the unending hills, the frequent elbows made by the N'fiss, now deeper and much fiercer than when we came, owing to the recent rains, which forced us to cross repeatedly and get wet every time; for the snow, which now lay thickly on the higher hills, though the sun was hot, still made the wind as cold as winter, and our clothes like paper on our backs. As we spurred along, calling, when a mule stumbled upon a narrow path above the

---

* Jaui is the word used in Arabic to designate Malays, Chinese, and the Eastern Mohammedans in general.

river, upon Sidi Bel Abbas, the saint of wayfarers, we racked
our brains to discover how we had been found out. The sun,
the winds, the want of food, the rain and dust of the road,
had by this time rendered us almost as dark as Moors. So
we were certain that we looked the character, for all the
passers-by saluted us with " Peace " and never turned their
heads to look at us, taking us evidently for travellers from
the Sus. My own idea was that the man we had taken from
Amsmiz had passed the word along the road. The others
thought, some one thing, some another, and so we talked
away, till at the last Mohammed el Hosein resolved the
matter, saying that Allah had not wished us to succeed.
Once more we crossed the rugged hill, where I had seen the
miserable donkey fallen on the road surrounded by his
wretched owners, or compeers. Again we stopped to rest the
animals under the oleander bushes by the stream; but this
time all dejected, hungry, and without refreshment, but a
pipe of kief, which Ali brought triumphantly out of a dirty
bag, and which we smoked by turns, sitting like Indians at
a council, waiting for the word.

Just about sundown we came to Taguaydirt-el-Bur, to
find it full; a sort of festival in process in the mosque; mules,
horses, donkeys, and armed men pervading all the place; a
whistling north wind blowing from the snow; nothing to
eat, and to be kept at least an hour sitting upon our beasts,
whilst the Sheikh read the letter we delivered from the Kaid.
At last a filthy negro came and called us in; whilst riding
through the dark and tortuous passage to the inner court,

my horse fell twice, once on a donkey and again upon a
mule: and the confusion, struggling, kicking, cursing, and
plunging in the dark, made me despair of getting through
without a broken bone. When we emerged the inner court
was full; men slept beside the donkeys, upon their pack-
saddles; and on the plaster benches running round the wall
beside them were their guns, their daggers, and their quarter
staffs; underneath their heads their money, valuables, corn
for their horses, or anything they chanced to value most.
Luckily none of them knew that we were Christians, but
thought most likely we were guests the Sheikh considered
worthy of his care; for no one looked at us, and we were
glad to pack into a room above an oil-press, smelling
abominably of oil, filthy, and with an opening in the roof
through which we saw the stars. Upon the doors and
shutters were some most curious concentric patterns cut deep
in the wood. They were not Arab, yet not like any European
work, nor yet were made by Jews; but they reminded me, as
the Kasbah at Kintafi had done, of the strange temples of
Palenque, the patterns on the Aztec Calendar, and the Sacri-
ficial Stone built into a side wall of the great church at
Mexico. Miserably we sat squatted before a pan of charcoal
trying to dry our clothes, and after some hours' waiting, a
man appeared with a dish of eggs fried in high-smelling
Argan oil, which we despatched at once, and tried to sleep;
but between the cold and the assaults of every kind of insect,
found it impossible, tired with the journey and hungry as
we were. All through the night a sort of Tenebræ seemed to

be going on in the neighbouring mosque; ghaitas were played, and every now and then a man broke out into a long high chant, the congregation answering him at stated intervals. Long before day I roused our people up, saddled, and picked our way through the courtyard where the travellers were still buried in sleep, then got upon the road just as the stars were paling, and a whistling wind blowing from off the snow, which chilled us to the bones. Mohammed el Hosein and Swani were afraid the Kaid might change his mind and send men after us to bring us back; so we pushed on as fast as we could go for about four hours, until the half-starved beasts began to tire, then walked a spell, and finally sat down to rest (this time without even a pipe of kief), and let the beasts try to pick up some grass.

Until that moment, so hurried had been our journey, I had not determined where to go; and after consultation, it being manifest we could not reach Morocco city by nightfall with our jaded beasts, and as I was not anxious to camp in the mountains another night without a chance of food, I determined to push on to Tamasluoght, the Sherif of which, Mulai el Haj, I knew and to whom I had sent a letter from Kintafi, asking him to agitate for our release. From where we were, the Zowia of Tamasluoght was distant at least eight hours' march, and if we wanted to arrive with light the time was short enough. So we pushed on; stumbling along the mountain paths, dismounting now and then to lead our animals, all silent and occasionally a little cross. The road branched to the east, and separated from the

Amsmiz path close to the place where we had seen a salt mine on our journey towards the Sus. Here for the first time we entered a real cedar forest; the trees not high, rarely exceeding thirty feet in height; but thick and growing closely so as in places to exclude the light. The trail led in and out between the trunks, and at times narrowed to about two feet in breadth, and went sheer down for seven or eight hundred feet. Unmindful of the proverb that "he who passes a bridge on horseback looks death in the face," * we passed without dismounting till Lutaif's saddle slipping almost precipitated him down the precipice. Dimly, and at an immense depth below, we saw the salt mine, looking like a glacier, and shining white as snow; the mode of working being to dig innumerable pits, from which the water ran into a sort of reservoir, where the salt lay heaped. Miserable donkeys crawled along the mountain paths, all grossly over-laden; and men passed stripped half-naked, but perspiring even in the chilly air, bent, like Swiss porters carrying tourists' boxes, under great loads of salt.

At last we began to leave the mountains, and emerged on to a broad and fertile plateau, much like Castille, and grow-ing the same crops of wheat, chick peas and vetches, with the resemblance increased by the mud-built houses, looking so like the hills that now and then we came upon a village almost before we could be sure if the brown heaps were irregularities of the ground, or houses, so exactly did the colours blend. In one of the small hamlets we bought bread

* "Quien à caballo pasa la puente, tiene la muerte ante la frente."

and fed our horses, and once again pushed on, till about
four o'clock, after winding up an interminable staircase of
rock, all of a sudden the plain of Morocco burst upon us; the
great city in the distance, and the Kutubieh standing like a
lighthouse of Islam, to guide the wanderer home. From
thence the descent was rapid; and in another hour we found
ourselves in a different climate, and again came to a country
where it had not rained for months, but was burnt up and
dry, with all the water-courses turned to brown holes, and
a dull shimmering in the still air broken but by the grass-
hoppers' shrill note. The next three hours were mortal, and
about sundown our animals reeled through the olive groves,
which for a mile or more surrounded the Zowia of Tamasluoght. We passed some water wheels, and drank at the first
of them; passed underneath the curious apparatus like a
switchback, supported on brick towers at intervals of a
hundred feet, by means of which water is taken to the
Sherif's house; rode through some sandy lanes, with houses
here and there; came to the gateway, and were met by
Mulai el Haj in person, who said he had not thought we
should have got out of Kintafi upon such easy terms.

No one could be more absolutely unlike our late lamented
host than Mulai el Haj, an Arab of the Arabs, descended
from the Prophet, but with little of the dignity that rank (in
the East) usually confers on its possessor. A man of peace
in fact, a semi-sacred character, and a sort of cross between
the Pope and a feudal baron of old time; rich, influential,
and occupying in the south the position held in the north by

the Sherifs of Wazan. When at last our diplomats became aware that the French had secured a footing in Wazan, and by their railway to Ain Sefra a point of descent on Fez and on Morocco city, it occurred to them to secure Mulai el Haj, as a set-off to the all-prevailing influence of the French throughout the north. So they induced our Government to extend British protection to the Sherif of Tamasluoght; but the protection has been so grudgingly extended as to be hardly of avail.

Into the ethics of the European occupation of Morocco I do not propose to go; but if at any time that occupation should occur, it is certain that Europe would not see us in Tangier. The French have thoroughly secured the north, and as the Germans will no doubt bid for some towns upon the coast, it might perhaps be advisable to take the south, and so control the Sus, secure Morocco city, and thus keep a way open ofr the Saharan trade, and opening Agadir, or some port on the Wad Nun, check French advance from St. Louis, Senegal, and Dakar, and their possessions in the south. I should prefer to see Morocco as it is, bad government and all, thinking but little as I do of the apotheosis of the bowler hat, and holding as an article of faith that national government is best for every land, from Ireland to the " vexed Bermoothes," and from thence to Timbuctoo.

Let us rob on in Europe as we have always done since first the Vikings sailed their long ships to plunder and to steal. England has never lagged behind in all adventures of this kind, so if there is a general scramble for Morocco, let us have our share. Statesmen can surely find reasons to justify us, and if they fail, we can sail in under the Jolly Roger, after the fashion of our ancestors upon the Spanish main. When we arrived at Tamasluoght, nothing was farther from my mind than politics, the advance of empires, ethics of conquest, Hinterlands, and things of that sort, on which the opinion of the first fool in the street is just as valuable as that of the politician fool, who stumbles out his halting speech to his bemused electors, who elect him for his likeness to themselves in density of head.

Mulai el Haj ushered us into a sort of kiosque, built in the style of the Alhambra, with a small court in front, in which grew cypresses and oranges, and would hear no word of anything before he had seated us on a comfortable divan in the recess of a window in his great chamber, which opened on a little garden where a fountain played, the water splashing on a marble basin, and the scent of flowers rising up to the room. After the Arab fashion he never left us for a moment,* and whilst a repast was being got ready, set tea before us, and then coffee, and, to pass by the interval, fresh bread and butter in a cracked though lordly dish. All was perfectly appointed in Arab style, the china French, the basin in which we washed our hands of brass most beautifully worked, and a black (uncomely) slave girl with heavy silver anklets handed round the cups. In her confusion at

---

* This fashion is still kept up amongst old-fashioned and provincial Spaniards, who, on the rare occasions on which they admit a stranger to their houses, never leave him till bedtime. It is called "to accompany the guest" (acompañar el huesped), and to omit it would be the height of ill-breeding.

the sight of Christians, instead of handing us a napkin she handed something which looked like the trimming of her drawers, and being rebuked in Shillah, retired in tears, and another equally ill-favoured damsel took her place. Her name I think was Johar, and the Sherif explained with gravity that negresses were as immodest as the hens, and that the slave had better have kept the trimming of her drawers to veil her face, though, as he said, what with the tattoo marks and negro features it contained little to tempt the eyes of a believing man. To this I gave a qualified assent, and through the medium of Lutaif guided the conversation on to more general grounds, on which a man, used from his youth, as Europeans are, to think all women angels in disguise, might comfortably join.

The Sherif having gone to the mosque for a brief interval, I walked into the courtyard to see my horse, and found him standing dejectedly and too worn out to eat; the mules exhausted, one of them lying down, with the men sitting beside them, too tired even to light a fire. Lutaif, who was not much accustomed to such trots, and who moreover had ridden a tired mule for most part of the day, had nearly fallen asleep during the tea and coffee and the long talk with the Sherif, now was a little rested, and strolled with me about the place.

The house was built in the usual Moorish style; with crenellated walls, flanking towers, and dome-shaped roofs. It had innumerable courts, a mosque, a women's wing, a granary, store-houses, baths, and everything in the first style of modern Arab taste. All round the houses stood cypresses, many of great age and height; over the mosque two or three palm-trees waved, and oranges and olives extended for some acres upon every side. There was a garden in the Eastern taste, where water trickled in a thousand little rills; canes fluttered, rustling like feathers in the air; jasmine and honeysuckle climbed up the azofaifa trees, sweet limes and lemons with pomegranates were dotted here and there; some periwinkles, but larger and paler than those grown in Europe, grew in the grass; here and there stood geranium bushes, straggling and run to seed. Over it all brooded the air of decadence, mixed with content, which makes an Eastern garden that of all others where a slothful and religious man should find heart-ease, and reading in some book of things beyond the power of intellect, leave them without solution, and sit still giving thanks for gardens as a true Christian should.

The inside of the house was decorated in a sort of pseudo-Persian style, with double doors all gilt, each with a little horse-shoe opening in the middle of it through which to pass, and on each side a formal tree like those upon the binding of a Persian manuscript. In fact, a modern replica of the Alhambra done without art, the gildings and the columns common, heavy and overcharged, and nothing really good except the iron gratings of the windows, and the tiles which, made today in Fez, seem to have but little deteriorated in glaze and colour from those left by the Moors in Spain.

The prayers over, the Sherif came back to keep us company, and sat till midnight talking of all sorts of things.

He appeared to think that we had great luck in not having been sent as prisoners to the Sultan's camp, for if we had been sent, we might have remained for weeks journeying about with the army before we were released. I was not sure if on the whole I was glad to have missed the experience, but as it would have been impossible to explain my reasons, I held my peace. Mulai el Haj, being an English protected subject, felt or assumed to feel great interest in English things. He asked most ceremoniously after our Grand Vizir (Lord Salisbury), was curious about the Queen, talked of the rising of the Afridis, and hoped that God would give the victory to our armies, a consummation which in his heart of hearts he could not really have wished, for no Mohammedan ever desires that Kaffirs shall triumph over "those of the faith." His questions answered with great detail, he launched into a disquisition on the present state of things in Morocco under the Vizirship of Ba Ahmed, "for the young Sultan rules but by his hand."

What he communicated may or may not have been his real opinion, for Arabs are apt to say that which they think will please their hearers; but situated as he was—that is, being a rich, powerful (and tolerably selfish) man—it may be that his words really conveyed his thoughts. "I have known tyrannies by Allah, and the late Sultan, Mulai el Hassan (may God have pardoned him), was not a lamb.

His father was something sterner, and their vizirs were apt, as vizirs always are, to fill their purses at the expense of powerful men. But since I first became a man, never a state of things so bad as now. Ba Ahmed passes all measure, grinds the faces of the poor, maintains the Sultan in a state of tutelage, taking his wives if so it pleases him, and sending them to his own house; he also sends the public money into France and puts it in their cursed banks, so that if the time for vengeance comes, he can escape and live upon it." Then looking round the room and moving up his cushions close to mine and laying one hand on my arm, he asked, " Do you catch all I say? " for poor Lutaif had almost dropped asleep. I understood him as he spoke slow and plainly, and he began. " Has England quite forgotten us, or does she sleep ? Time was when she was not wont to wait long when a country lay open as does Morocco to her power." I thought he under-stood our policy and what has made us so beloved by every-one, so nodded, and he went on. " I can insure the southern chiefs of all the tribes. I, Mulai el Haj, who speak, but who through the intrigues of Ba Ahmed dare not leave his house; twice have they fired at me walking in my own garden, once when on my horse almost a mile outside, and even now I know that men are waiting for me did I go out. But still I can insure the southern chiefs from here to Taseruelt, westward to Tuats, to Tafilet, and through the Atlas; the dwellers in the plain around the city of Morocco all either send emissaries or visit me by night. I tell you, even the wildest of those who think a Christian is not a human being are so hard pressed, that if the English came they would meet them on the road and pour out milk.* God instituted government, as he made moon and sun, set one by day to shine, the other to guide wanderers by night, and as he set the stars in the blue heavens, so he set Kaids and Governors, Sheikhs, Mokadems, Cadis, and all the hierarchy of rulers, each in his place to rule mankind. The tail can never be so honourable as is the head, and whilst men still exist they must be ruled, ruled justly; but this Ba Ahmed knows no justice in his heart." Passion o' me, I thought, he is no socialist, nor for that matter is the poorest Arab, all thinking that authority came straight from God. Then he continued, "For two years I have never dared to go to Morocco city, though but three hours away. My houses there are ruined. Ba Ahmed has taken one of them, and the other stands open a prey to wind and rain, with all the woodwork torn away and burnt. Your 'Bashador' two years ago advised me to make friends if possible with the Vizir. Therefore I bought a female slave, not to go empty-handed for I knew well that those in office always expect a present from all those they see. Three hundred dollars did I pay for the girl; not that she was a houri whom the sheep would lift their heads to look at if she walked across a field, but passable, and fitting to present to a man who like Ba Ahmed filled the office of Vizir.

" And so I went to see my enemy. Why he should be my enemy I do not know, except that I am rich and am known

* To pour out a libation of milk on the road is a sign of welcome in Morocco.

to favour England, which he detests; but as it is, may Allah put his mercy some day into the hearts of flint. The girl I placed upon a mule, and taking with me twenty or thirty well-armed men, rode to Morocco city. Arrived before his door, I had to wait two hours, I a Sherif, to wait two hours to please the mulatto dog; and then the slave who led me to him brought me through many passages and left me standing outside a half-opened door, whilst he went in. Long did I stand there, and being angry at the indignities that I was passing through, forebore to listen till at last I heard Ba Ahmed say to his slave, 'Where is the man ?' The negro answered, 'He is come,' and then Ba Ahmed angrily replied, 'Did you not take him where I told you ?' (i. e. to be killed), and the slave excused himself, saying: 'I thought you wanted him in here.' After a moment I heard Ba Ahmed say: 'I told you plainly to take him to be killed, but God has spared him, let him come in.' Then the slave threw the door open, and I advanced, was kindly welcomed, sat and drank tea, thinking each cup was poisoned, but made no sign, knowing that Allah, who had spared my life a moment previously, could turn the poison into sugar, if he willed it so.

"We sat and talked, and then I gave the slave girl to him, and she was led away into his house. I took my leave, and he with courteous words urged on me return and visit him; but who that once escapes from the lion's den places his head again beneath his paw ? God has indeed been gracious to me, praise his name, The One." As he stopped

I thought, indeed, who that would pity a snake-charmer who is bitten by a snake, or any one who cometh near wild beasts? I did not say so, but confined myself to praises of his prudence and of the special providence which had saved his life.

Then I enquired if he had got my letter from Kintafi, saying jocularly that I had looked for him to come to my assistance waving a Union Jack. To my surprise he denied all knowledge of the letter, and said he had only heard of my captivity by accident. On going to Morocco city, the day after, I found a special messenger had been sent off to him bearing the letter; but after all blessed are they who expect little, they shall be satisfied.

At midnight, and when Lutaif had long subsided into slumber on the divan, leaving me to puzzle out the worthy gentleman's discourse as best I could, he took his leave after urging me warmly to bear his compliments to our Grand Vizir. I pitilessly woke up Lutaif and had a consultation with him, what we should do, it being quite impossible for the animals to return to Mogador without some rest, and equally impossible to remain at Tamasluoght, where the Sherif would let us pay for nothing, either for ourselves or for our beasts. Morocco city was but three hours off, and as I expected letters there, we arranged to stop two or three days and feed our animals, and then push on across the plains for Mogador. Next morning saw us in the saddle by ten o'clock; and after a courteous exchange of compliments with the Sherif, and renewed entreaties on his part to put his view of the state of affairs in Morocco before the Grand Vizir on my return, we took our leave.

Thus for the second time I passed a night at Tamasluoght, having once, four years before, been there with Mr. Harris* on the occasion when the Sherif was agitating for British protection; and as I saw him standing before his door and bidding us good-bye, a courteous, prosperous, saintly Moorish gentleman, it came back to my mind that I had said to Mr. Harris that his protection, as far as English interests were concerned, was thrown away.

If we protect at all, except from pure philanthropy, we should protect lean, sunburnt sheikhs, who pass their lives on horseback, and at whose call spears and long guns rise from the desert like frozen reeds stick through the ice in ponds in winter time; and if that sort is not available, why those of the same kidney as the man in whose house I slept next night, Abu Beckr el Ghanjaui, prince of all Moorish intriguers and diplomatists; one who will stick to any cause through thick and thin if it is worth his while, and who, I fancy, if he chose to speak all that he knows of British policy in Morocco, could make some diplomatic folk get up and howl, or send them snorting like an Indian pony about the Foreign Office.

* Walter B. Harris, author of " Tafilet," and many works on Morocco, and one of the few Europeans who really know the country and the Moors.

Chapter XII

ONCE through the olive groves of Tamasluoght, the city of Yusuf-ibn-Tachfin* lay glistening on the plain, almost hull down on the horizon. Above the forests of tall date palms which fringe the town, the tall mosque towers rose, the Kutubieh and the minaret of Sidi Bel Abbas high above the rest. From the green gardens of the Aguedal the enormous stone-built pile of the Sultan's palace, all ornamented with fine marbles brought from Italy and Spain, towered like a desert-built Gibraltar over the level plain. Across the sea-like surface of the steppe long trains of camels, mules, and men on foot crawled, looking like streams of ants converging on a giant ant-hill, whilst in the distance the huge wall-like Atlas towering up, walled the flat country in, as the volcanoes seem to cut off Mexico from the world outside. The situation of Morocco city much resembles that of Mexico, which had a pseudo-Oriental look, the flat-roofed houses and the palm trees completing the effect.

* Morocco city was founded by Yusuf-ibn-Tachfin in 1072, on the site of the ancient Martok, and near where some say the Romans had a city with the strange name of Bocanum Homerum, sounding like nothing Roman and perhaps an attempt of the Romans to write some Berber word. Yusuf-ibn-Tachfin was the first prince of the Almoravides who invaded Spain and overthrew the forces of Alfonso VI of Castile at the great battle of Alarcos, and reigned over Southern Spain and Africa until the Almohades broke their power.

A hot three hours, kicking our tired beasts along, brought us outside the city walls, and passing underneath the gate, which zig-zags like an old Scottish bridge, we emerged into the sandy lanes running between orange gardens, which form a kind of suburb of the town, and where the Soudanese, the men from Draa and the Wad Nun, do mostly congregate. No one would ever think, from the aspect of the lanes, unpaved and broken into holes by winter rains, that he was actually inside a city which is supposed to cover almost as much ground as Paris. It took us almost three-quarters of an hour to ride from the outside walls to the centre of the town. We passed through narrow lanes where camels jammed us almost to the wall; along the foot-paths beggars sat and showed their sores; dogs, yellow, ulcerous and wild as jackals, skulked between our horses' legs. At last we came out on an open space under the tower of the Kutubieh, in which square a sort of market was in progress, and a ring of interested spectators sat, crouched, and stood, intent upon a story-teller's tale. I sat a moment listening on my horse, and heard enough to learn the story was after the style of the Arabian Nights, but quite unbowlderized and suitable for Oriental taste.

A certain prince admired a beauteous dame, but an old Sultan (always the wicked baronet of Eastern tales) desired her for his harem, and engaged a certain witch, of whom there were great store throughout his territory, to cast a spell upon the prince, so that the lady should fall into a dislike of him. He, on his part, resorted to a wizard who stirred the ladies of the Sultan's harem up to play strange pranks and turn the palace upside down, let young men in o'nights, stay out themselves too late, and generally comport themselves in a discreditable way. A faithful slave at last made all things right, and after a most realistic love scene the prince and princess were married and lived happy ever after; or, as the story-teller, a sad moralizing wag, remarked, until the prince should take another wife. Humanity, when crushed together in the heat, either in London ball-rooms or in waste places in Morocco city, sends up a perfume which makes one regret that the cynical contriver of the world endowed us with a nose. Therefore I waited but a little and rode on, turning occasionally to take a look at the great mosque and the tall dusty tower. The outside of the mosque, the name of which in Arabic means Mosque of the Books, from the word Kitab, a book, is not imposing. What it is inside I believe no Christian knows. Had I that moment, dressed as I was, sunburned and dirty, got off and entered it, I might have seen, but the thought did not cross my mind, and afterwards, when known for a European, it might have cost my life. The tower springs straight from the sandy square as the Giralda rises from the level of the street in Seville. One man built both, so runs tradition, and certainly the Kutubieh tower today reminds one greatly of the description of the Giralda when San Fernando drove out the Moorish king of Seville, and planted the banner with the Castles of Castille above the town. The same gilt globes, of which the Spanish speak, are on the Moorish

tower, and the same little cupola which the Christians took
away in Seville, replacing it by a renaissance " flèche," upon
which stands the towering figure cast by Bartolomé Morel.
The tower, almost three hundred feet in height, is built of
dark-red stone, with the alterning raised and sunk patterns
(called in Spanish Ajaracas) cut deeply or standing boldly
out from the solid masonry. At one time tile work filled
most of the patterns, or was embroidered round the edges of
the windows, but neglect and time have made most of it
drop away. Still, just below the parapet runs a broad band,
which from the square appears to be full four feet broad,
of the most wonderful black and green iridescent tiles I
ever saw. When Fabir, who, tradition says, built it for the
Sultan El Mansur, and it stood glorious, adorned with tiles
like those which still remain, the gilding fresh upon the
great brass balls, even the mosque at Cordoba itself could
not have been more glorious, and El Mansur could not have
easily foreseen that on his lonely tomb under the palm
trees, beside the river at Rabat, goats would browse and
shepherds play their pipes. Allah, Jehovah, all the Gods are
alike unmindful of their worshippers, who made and gave
them fame; what more may the contrivers of the Crystal
Palace and the gasometers at Battersea expect, when they
have had their day ? Medina, Mellah, Kaiserieh, Sidi Bel
Abbas, the tomb of Mulai Abdul Aziz, all have been described so many times and by such serious and painstaking
writers, who have apparently measured, re-measured, and
calculated the cubic capacity of every building in Morocco

city, that it would have been a work of supererogation on my part to have laid a measuring tape once more on any of them.

Morocco city struck me, and has always done so, for I have been there twice, as the best example of a purely African city I have seen. Fez has the mixture of Spanish blood in its inhabitants which the expulsed from Malaga, Granada, and from all the Andalos, brought and disseminated. In the high houses, which make the streets like sewers to walk in, you hear men play the lute, and women sing the Malagueña, Caña and the Rondeña as in mountain towns in Spain. Quite half the population have fair hair, some pale blue eyes, and their fanaticism is born of ancient persecution by the fanatic Christians of Spain. In every house, in every mosque, in almost every saint's tomb is fine tile work, stone and wood carving, the eaves especially being often as richly decorated as they had been Venetian and not African. The streets are thronged, men move quickly through them and the whole place is redolent of aristocracy, of a great religious class, in fact has all the air of what in Europe we call a capital.

Morocco city is purely African; negroes abound; the streets are never full, even in the kaiserieh * you can make your way about. With the exception of the Kutubieh Tower, and some fine fountains, notably that with the inscription "Drink and admire" (*Shrab-u-Schuf*) inscribed upon it,

---

* Kaiserieh, the bazaar. Literally silk market. The word is preserved in Spanish under the form of Alcaiceria.

and the fine gate of the Kasbah of the best period of Moorish work, there is no architecture. Sand, sand, and more sand in almost every street, in the vast open spaces, in the long winding narrow lanes, outside the walls up to the city gates; sand in your hair, your clothes, the coats of animals. Streets, streets, and still more streets of houses in decay. Yellow adobe walls, dazzling white roofs and dense metallic semitropical vegetation shrouding the heaps of yellowish decaying masonry. No noise, the footfalls of the mules and camels falling into the sand as rain falls into the sea, with a soft swishing sound.

The people all are African, men from the Draa, the Sus, the Sahara, Wad Nun and the mysterious sandy steppes below Cape Bojador. Arabs are quite in the minority, and the fine types and full grey beards of aged Sheikhs one sees so frequently in Fez exchanged for the spare Saharowi type, or the shaved lip and cheeks and pointed chin tufts of the Berber race. Tom-tom and gimbry are their chief instruments, together with the Moorish flute, ear-piercing and encouraging to horses, who when they hear its shriek, step proudly, arching their necks and moving sideways down the streets as if they liked the sound. Their songs are African, the interval so strange, and the rhythm so unlike that of all European music, as at first hearing to be almost unintelligible; but which at last grow on one until one likes them and endeavours to repeat their tunes. Hardly an aristocratic family lives in the place, and few Sherifs, the richer of the population being traders with the Sahara.

A city of vast distances, immense perspectives, great desolate squares, of gardens miles in length, a place in which you want a mule to ride about, for to attempt to labour through the sand on foot would be a purgatory. And yet a place which grows upon you, the sound of water ever in your ears, the narrow streets arched over all with grape vines; mouth of the Sahara, city of Yusuf-ibn-Tachfin, town circled in with mountains, plain girt, sun beaten, wind swept, ruinous, wearisome, and mournful in the sad sunlight which enshrouds its mouldering walls.

Fez and Rabat, Sefrou, Salee and Mogador with Tetuan, Larache, Dar-el-Baida and the rest may have more trade, more art, more beauty, population, importance, industry, rank, faith, architecture, or what you will; but none of them enter into your soul as does this heap of ruins, this sandheap, desert town, metropolis of the fantastic world which stretches from its walls across the mountains through the oases of the Sahara; and which for aught I know may some day have its railway station, public houses, Salvation Army barracks, and its people have their eyes opened, as were those of Adam and of Eve, and veil their nakedness in mackintoshes. Through streets and open spaces, past mosque doors, with glimpses of the worshippers at prayers seated upon the floor, or lounging in the inner courts, through streets arched in with vines, the trellis work so low that upon horseback one had to bend one's head in passing, and at the side door of the missionary's house (Dar-Ebikouros) I got off, and, sending up a boy, was met by Mr. Nairn, who for a moment did not recognize me dressed in the Moorish clothes. There, as upon my first visit to Morocco city, I received a hospitable and courteous welcome. Long we sat talking of our captivity. I learned about the hurried visit of the Oudad with letters, his departure without a word, and found that no one had expected us so soon. Mr. Nairn, who spoke both Arabic and Shillah well, had passed on one occasion close to Kintafi; but, not having been near the castle, was not recognized, but like myself had been unable to push on to Tarudant. This in a measure consoled me for my failure, as Mr. Nairn had lived long in the country, spoke the language well, and with his dark complexion and black hair, dressed in the Moorish clothes, must have looked exactly like a Berber mountaineer. After a welcome and most necessary bath I left his hospitable house and rode to Sid Abu Beckr's, almost the only man in Morocco city from whom it is possible for a European to get a house. And, as I rode, I mused upon the mystery of faith, and marvelled to see the honest single-hearted missionary still with the cross upon his shoulders, ploughing the stony vineyard of the Moorish heart, quite as contentedly and just as hopefully as he had done four years ago. Yet, not a ray of hope, without a convert or a chance of making one, and still contented, hoping for the time when he should see the fruit of his hard work. Crowds thronged his courtyard in the morning to get medicines, and I fancy as he dispensed his drugs, in the goodness of his heart he tried to do all that was in his power to lead his patients to what he thought the truth. Women in numbers came, not for the medicines, so much as for the bottles, which they valued highly to keep oil for cooking in, throwing the medicine carefully away; but cherishing the flask and bearing it about them always, slung in a little case. Bottles may yet save souls when preaching fails, for women who receive them may so work upon their husbands' hearts, that by degrees, from the errors of Mohammed and the mere two ounce phial, they rise to the imperial pint and Christianity; and so societies at home should send more bottles out, artfully coloured and with the necks fashioned to hold a string, so that if bibles prove of no effect, bottles may yet prevail.

Not that I mean to undervalue missionaries; they have their uses, but in a different way from that in which perchance they think themselves. What they can do is to set forth, in countries like Morocco, that they are not mere merchants trying to deceive all those with whom they deal. So in Morocco city Mr. Nairn and his wife, and the young men and women of his household, have the respect of all the Moors for the pureness of their life, and their untiring kindness to the poor. The educated Moors see that they are like their own religious sects—that is, their minds are fixed, not upon gain, but prayer, and in the East madness and holiness are held akin, and both, as being sent from heaven, are respected.

In a few hundred yards I left all holiness behind, and getting off at Abu Beckr's door, found that astute and clever politician seated as usual at the receipt of custom, counting some money which he had just received, a pistol by his side and a large iron box wide open by him in which to store the gold.

Sid Abu Beckr el Ghanjaui is known from the Atlas to the Riff, and from the Sahara to Mogador, feared and disliked, and yet respected, for the Moor above all other things respects success. Not that, by any means, he thinks the less of those who fail. Success and failure are both sent by God, they were ordained (Mektub), and the mere man is but the instrument of Allah's will. For the last thirty years Sid Abu Beckr has been the British agent in Morocco. During that time he has made many enemies, as any man of ability placed in his position could not fail to do. Of obscure origin and deeply tinged with negro blood, he is, perhaps, today the richest and the ablest man in the whole country. Few men in any land have been the victims of more calumnies, but, on the other hand, few men have had more friends. To some, a slave dealer, a traitor, spy, and perjured sycophant; to others, a true friend of England, a man who has suffered much for his devotion to her cause. To me, a clever, scheming politician, who has known the right way in which to play upon the weaknesses of a long series of Ambassadors. A friend to England without doubt, as thrasher is to sword-fish when they attack a whale. A true " faux maigre "; a thin yet flabby man, scant bearded like a eunuch, reedy voiced, and in complexion atrabilious; his shoulders bent, eyes with spots on the yellowish whites, and pupils like a cat's; slight nervous hands, persuasive manners, and, in fact, one who impresses you at first sight as a keen intellect confined in a mean envelope; but yet not despicable.

Not what is called an educated Moor, still less of the Moorish upper classes by his birth, he yet today has as much power in Morocco as any man outside the circle of the court. He says he speaks no English, though I think he understands it, but he takes care to conceal such knowledge of it as he may have, so that in speaking to him, those who speak no Arabic, speaking through an interpreter, may give him ample time for the consideration of every word he says. Never for a moment to be caught off his guard or disconcerted, for the story goes that once, during an ambassadorial visit, a slave dealer came with some merchandise, and that, in answer to the query of the Ambassador if those were slaves, Beckr replied they were, but, as he was an English subject, when they entered through his doors they became free at once.

For some reason, not perfectly explained, the Non-conformist conscience, some years ago, was greatly exercised about the man. All that was infamous was put down to his charge. He was a slave-dealer, a brothel keeper, I think a murderer, and, of course, an adulterer, that being the crime the "Conscience Bearers" detested most, being shut out from all participation by the exigencies of their life. All went on merrily as things are apt to go in England when the accused is a good long way off: questions were asked in Parliament as to why England countenanced such a man as Abu Beckr in a position of high confidence. Ministers answered what was put into their heads, having no inkling of who Abu Beckr was, or what had raised the storm. As there was after all no money in the matter, the questioners gradually tailed off. Then Abu Beckr brought an action in the court in Gibraltar, and explained how it was he had been charged with the commission of so many crimes. It turned out that being a Moor he certainly had slaves, and even bought and sold them as we do horses; but as such was the everyday custom of the land, and he, when he took British protection, had not become a Wesleyan Methodist, where was the harm in it? As to the second charge, he had, of course, four wives, and no doubt many women in his house, but, as he pertinently said, that his religion allowed him, and as far as he knew yet, he had not changed his faith. So he triumphantly floored his antagonists, got damages, received eventually a silver tea service from the British Government, and retired to his home to laugh at every one concerned.

Abu Beckr, though he knew I had arrived in town, most likely took me for a Moor from Fez, between whom and the people of Morocco city little love is lost, for somewhat roughly he asked me what I wanted, and did not offer me a seat. I listened to him for a moment, and watched his cunning diplomatic smile, as he looked at me from the corner of his eyes to find out who I was. Then I said in English, " Good morning, Sidi Beckr "; and he laughed and said he had known me all the time, but wanted to see what kind of Arabic I spoke. As he had only seen me once before some years ago, and dressed in European clothes, I only smiled, and said my Arabic was worse than ever, and that I wanted him to lend me some house in which to rest my servants and myself.

A man was brought to accompany us to an empty house hard by; he bore a monstrous key, and after leading us through several narrow streets, stopped at a brand new house, and throwing the door wide open said it was fit for any king, and that he generally received five dollars upon Abu Beckr's account for opening the door. I told him instantly that I should mention to his master what he said, then took the key out of his hand, gave him a dollar, and asked Lutaif to tell him to what place the Koran condemns all those who palter with the truth. The usual scrubbing and swilling out took place, which always has to be undertaken before it is possible to occupy an empty Moorish house. Once swilled and dried, we installed our scanty property, sent out for food, put up our horses in a fondak not far off, and fell asleep upon the floor. Upon awakening we found the dinner beside us, and a negro squatting, and patiently watching till we should awake.

During the interval, Swani and Mohammed el Hosein had both gone to the bath, and then I fancy, after the fashion of all sailors and muleteers after a voyage or trip, gone on the spree, for in the morning they appeared like Mr. Henley's " rakehell cat," looking a little draggled, and

the worse for wear, and swearing that they had not touched
a drop of drink. The patient Ali never stirred away, being,
as he said, rather afraid to venture out alone amongst the
people in the crowded streets. After a journey in Morocco
the men always ask for new shoes, so to show my disgust
at the immoral conduct of the others, I took Ali out and
made him happy with a pair of evil-smelling yellow leather
shoes, adding a pair of gorgeously embroidered orange-
coloured slippers for his House.

A dirty little negro boy came to inform us that Sidi Abu
Beckr expected us to dine that afternoon at two o'clock,
the fashionable hour in the Sherifian capital.

During the interval I walked about the streets, pleased
that the Moorish clothes relieved me from the attentions
to which I had been subjected on my former visit to the
place. Through the interminable streets I strolled, past
ruined fountains and the doors of mosques half opened,
from whose interiors came a sound of prayer, as from
the beehive comes the murmured prayer of bees. Long
trains of camels pressed me into corners to escape their
snakelike heads, and suddenly, and without consciousness
of how I got there, I found myself in a remembered spot.
A little alley paved with cobble-stones and bordered on
each side with open shops, in which sat squatted white
figures working hand looms which filled the alley with
their clack. At the end an archway with a wooden gate
hanging ajar. I entered it with a strange feeling of possession, and found everything familiar; the tank, with edges

of red stucco work, the Azofaifa* trees, the bordering hedge
of myrtle, the white datura, the curiously cut semi-Italian
flower beds, in which grew marjoram and thyme, the open
baldachino at the end, under whose leaky roof a friend
and I had spread our rugs and spent ten happy days four
years ago, smoking, lounging about, and talking endlessly
of nothing, as only friends can talk, was all unchanged.
What are four years of inattention beside the perennial
decay of all things Eastern; the winter rain and summer
sun had scarcely put an extra stain or two upon the plaster
work, the door was made to last for ages, and the trees
had but become a little more luxuriant; so, sitting down,
I smoked and fell a-musing on the time when in Morocco
all would be changed, and places like the Riad† el Hamri,
where I then sat, exist no more. Railway engines (praise
him who giveth wisdom to mankind) would puff and
snort, men hurry to and fro, tramways and bicycles make
life full and more glorious; women unveiled would sell
themselves after the Christian way for drink and gold; men
lie drunk on holidays to show their freedom from debasing
superstition; all would be changed; the scent of camels'
dung give place to that of coal dust; and perhaps Allah,
after regarding with complacency the work of man, would
rest contented as he did in Eden when his first masterpieces
forced his hand.

* Azofaifa is the jujube tree (*Zizyphus jujuba*). The Spanish word azofaifa is
derived from the Arabic zofaif.

† Riad means a garden. In Andalusia it is still used under the form of Arriate.

Sid Abu Beckr met us in the courtyard of his house, dressed in a light green robe, spotless white haik,* new yellow slippers, with a large rosary in his hand, although the oldest citizen of Marakesh had never seen him pray. Leading me courteously by the hand into an upper chamber looking upon a courtyard, and decorated in the purest modern Alhambraesque, he seated me, Lutaif, and then himself, on cushions, and we ate solidly for a full hour, until the welcome sight of tea, served in small gold cups, announced our sufferings were at an end.

Nothing would get out of his head at first that I was not an agent of the Globe Venture Syndicate, or had a mission from the British Government to try and establish some sort of undertaking with the chiefs of Sus. His apprehensions set at rest, or at the least so he pretended, he, too, began to give his views upon the state of current politics. All that the Sherif of Tamasluoght had said he quite corroborated, but with the difference that what he said was the opinion of a quick-witted, clever man, and therefore, to my mind, to some extent brought less conviction than the mere wanderings of the artless " saint."

Most Arabs of the richer class are quite incapable of seeing anything but from a personal standpoint, and thus it is that, bit by bit, their national power has fallen into decay. In Spain, Damascus, Tunis, and the Morocco of to-day chiefs have arisen, and their sole idea has been to push their individual fortunes. Except the Emir Abd-el-Kader,

* The haik is supposed to be the Roman Toga, and it is certainly very like it.

none have had even a glimpse of trying to restore the Arab power.

So Abu Beckr regaled us with stories of Ba Ahmed's villainy, his own position and personal insecurity, and assured me that the end was near, and that if England still stood hesitating, France would step in, and lastly, getting for a moment out of his egotism, said, " What avails it that I am so rich when my favourite son died of small-pox only three weeks ago, and I have no one to whom to leave my wealth?" Then he got up and said he had never showed his wealth to anyone, but as he found me so sympathetic to his views (I not having said a word), he would show me all his treasures and his house. Accordingly, we followed him into an interior court, on one side of which a door with about twenty padlocks stood.

Sidi Abu Beckr having assured himself that no one but ourselves was looking on, began to clash and bang amongst the locks and bars, piling the padlocks on the ground, unloosening chains, and making as much noise as a battle of armour-plated knights must have produced in the never-to-be-forgotten days of chivalry, when, in their mail, the scrofulous champions tapped on one another's shields. At last the door swung open and disclosed a room packed full of boxes, silver dessert services, china of all sorts, lamps, clocks, and every kind of miscellaneous wealth collected in the long course of an honourable career by strict attention to all economic laws. Boxes were there of iron painted in colours, made in Holland and in Spain, standing be-

side hide cases, teak chests, and old portmanteaux, Sara-
toga trunks and safes, cash boxes, sea chests, and packing
cases, and all apparently stuffed to the very lids. With
pardonable pride, and with a flush stealing upon his parchment cheek, he did obeisance to his gods. " This is all silver
Spanish dollars, this Sultan Hassan's coins, this packing
case is jewellery. I had it all in pledge. This, silver in the
bar, and these hide bags are gold dust, but the king of all,"
and here he touched an antiquated safe, " is el d'hab (gold),
chiefly in sovereigns. Yes, I know the waste of interest, but
do not think you see half of all my wealth; the bulk of
it is safely placed in England—consols, I think they call
it—safe, but small returns, and smaller since the accursed
Goschen lowered the rate. This that you see I keep beside
me, partly from caprice, for el d'hab has always been my
passion, passing the love of women, horses, or of anything
that God has made to ease the life of man." As he spoke
he patted the safe with his slippered feet, and looked as if
he knew a special providence watched over him; but knowing did not truckle to the power, but rather took it, as a
tribute to himself and his ability, as being well assured that
providence is always with the strong. As I looked at him,
proud of his wealth, his cunning and his good fortune,
it seemed as if our nation, with its power, its riches, and
its insensibility, was fitly represented by the worthy man,
who, from a camel driver's state, by the sheer force of industry and thrift, had made his fortune.

Blessed are those who rise, to them the world is pleasant

and well ordered, all things are right, and virtue is rewarded (in themselves); thrice blessed are the pachyder-
matous of heart, the deaf of soul, the invertebrate, insen-
sible, the unimaginative; nothing can injure, nothing
wound them; nature's injustice, man's ineptitude, fortune's
black joking, leaves them as untouched as a blind cart-
horse, who, in struggling up a hill, sets his sharp, calkered
foot upon a mouse. But as mankind, in strangeness and
variety of mind, is quite incomprehensible, Sidi Abu Beckr
had but hardly locked the door upon his hoards than he
set to bewail his sonless state. " Allah has given me children
(so he said), sons, grown to men's estate, but all unprofit-
able, idle and profligate, and of those who spend their time
in folly, all but one, a boy to whom I hoped to leave my
wealth. All that you see was to be his, the silver, gold, this
house, my lands, investments in your country, all I have,
and as I thought that Abu Beckr had not lived in vain, the
pestilence fell on my house and left me poorer than when
I was a camel driver; but it was written; God the most
merciful, the compassionate, the inscrutable, he alone giveth
life, and sends his death to men. Come let me show you
where my blessing died."

Lutaif, who, since the word smallpox had first been
mentioned, had held his handkerchief up to his nose, now
for the first time in the journey almost broke into revolt.
Pulling my " selham," he whispered, " Let us go, the in-
fectious microbes stand on no ceremony," and, in fact,
wished to retreat at once. Not from superior bravery, but

because I knew if there was mischief it was already done, I laughed at all his fears, told him to trust providence, as a good man should do, and that for microbes, probably by this time a new school of scientific men said they were non-existent, and put down all diseases to some other cause. Seeing he got no sympathy from me and perhaps after prayer (prayers, idle prayers) he followed, and Abu Beckr conducted us to a room and opened wide the door. Within, piled up upon the floor were heaps of rugs, and evidently no window had been open for a month. " Here (he said) died my son, and nothing has been touched since he was buried; upon this very rug he breathed his last." And as he spoke, he lifted up a carpet from the Sahara, woven in blue and red, and moved it to and fro so that the microbes, if there were any, must have had fair play to do their work. Lutaif turned pale, and once more pleaded to be off; but Abu Beckr was inexorable and putting down the rug, led us again through a long passage and opening yet another door, said, " Here the mother died, and nothing has been altered since her death." At last Lutaif grew desperate and seeing there was no escape began to grow at ease and followed us through countless passages, peeping into rooms in each of which some member of the family had died.

On our return to the alcove, where tea awaited us, we passed a figure swathed in white which turned aside to let us pass, its face against the wall. Abu Beckr took it by the hand and introduced it as the mother of the dead boy's mother, and a hand wrapped in a corner of a veil stole out, and for a moment just touched mine, but all the time the eyes were fixed upon the ground, after the style of Arab manners, which ordains that a woman must not look a strange man in the face. Something was mumbled in the nature of a complimentary phrase and then the sheeted figure slipped mysteriously away, but no doubt turned to look at the strange animals through some grating in the wall.

Lutaif explained that the honour done us was remarkable, and that no doubt Abu Beckr had stopped her to show how free from prejudice he was. The visit ended with more tea and a long talk, in which again I was assured the end was near, and that the Grand Vizir had made the life of every one intolerable, ruined the land, rendered the Sultan despicable, and that all educated men longed for the advent of some European power. Having heard the same tale from the Sherif of Tamasluoght, and as I knew both he and Abu Beckr were rich men and above all things feared attacks upon their wealth, I mentally resolved to talk the matter over with some Arab of the old school and hear his views.

Standing before the door, we found a soldier with a lantern waiting to see us home. We bid farewell to Abu Beckr, watching him stand beneath the archway of his house in his green robe and white burnous, looking a figure out of the pages of the "Arabian Nights"; but with a scheming brain and subtle mind, able to hold his own with trained diplomatists and to defeat them with the natural craft implanted in him by a wise providence which arms the weak with lies, makes the strong brutal, and is apparently content to watch the struggle, after the fashion of an English tourist gloating upon a Spanish bull fight on a fine Sunday afternoon.

After a two days' stay, with mules and horses well fed, I left the city, passed through the battlemented gates, and saw the walls, the gardens, palm trees, towers, and last of all the Kutubieh sink out of sight, then set my face westward to cross the hundred and thirty miles of stony plain which stretches almost to Mogador. After the first day's ride, our animals showed signs of giving out, the starving at Thelata-el-Jacoub having reduced them so much in condition that the two days' rest had done but little good. Those who have ridden tired horses, through stony wastes heated to boiling point by the sun's rays, and without chance of finding water on the road, can estimate the pleasure of our ride. On the third day, just about noon, and after toiling painfully at a slow walk through interminable fields of stone, we reached the Zowia of Sidi-el-Mokhtar, where a portion of the tribe Ulad-el-Bousbaa (the sons of Lions) were established, the other portions of the tribe being respectively in Algeria and in the Sahara.

Nothing could possibly have been more desert-like than the surroundings of this tribe. On every side stones, stones, and still more stones. For vegetation thorny Zizyphus, hard wiry grass, stunted euphorbiaceæ, with colocynths growing here and there between the stones; a palm-tree at the well, and a gnarled sandarac tree, looking as if it had never known a shower, with the leaves as hard as pine needles, stood like a sentinel defying thirst. The Zowia itself seemed even more Eastern than most buildings of its kind, the walls merely baked mud, the towers but hardly overtopping them, and all the animals of a superior class to those raised by the ordinary Moors. The tribe ranks high for fighting qualities as riders, shots, and swordsmen, and though small, can hold its own against all comers. The women, taller and thinner than the women of the Moors, were dressed in blue, after the desert fashion, and in procession walked to the well, with each one carrying an amphora upon her head.

The Sheikh, a Saharan Sherif, by name Mulai Othmar, was the best specimen of a highcaste Arab I had ever seen. Tall and broad-shouldered, lean and tanned by the sun to a fine tinge of old mahogany, grave and reserved, but courteous. Arabia itself could have produced no finer type of man. He asked us to dismount, put corn before our beasts, and sent a dish of couscousou for ourselves, sat and conversed, drank tea, but would not smoke, saying to smoke was shameful, but all the same he was not able to forbear to ask if tobacco was as strong as kief. Strength being the first object to an Arab, he not unnaturally believed that as our powder far exceeded theirs, so did tobacco far exceed their kief in strength, and seemed a little disappointed when I told him kief was stronger far than any cigarette.

During the two hot hours of noon, whilst the south wind blew like a blast out of a furnace, we squatted underneath the Zowia wall, shifting about to dodge the sun, as it moved westwards, and the Sherif stayed talking with us, as befits a man of blood. Though simply dressed in clean white clothes, he looked a prince, and all our men saluted him with ten times more respect than they had used towards either Abu Beckr or the Sherif of Tamasluoght. Much he imparted of the Sahara, its lore, traditions, told of the "wind drinkers,"* who, in the ostrich hunts, carry their masters a hundred miles a day. Much did he tell us of the Tuaregs, who, being Berbers, are at constant feud with all the Arab tribes. In the Sahara no money circulated, but a good mare was bought for forty camels or two hundred sheep; guns came from Dakar, or St. Louis Senegal, and were called either "Francis," or else "Mocatta," though he could give no explanation of the latter term. Beyond St. Louis was situate the mysterious Ben Joul,† to which place he said the Inglis came, a nation at eternal enmity with the Francis, fair in complexion, and addicted to strong drink, but pleasant to deal with, and in business having but one word.

Morocco seemed to him too green and overgrown with vegetation, so that a man grew dazzled when he looked at

---

* "Wind drinkers," Shrab-er-Reh in Arabic, is the term applied in the Sahara to the best breed of horses.

† Geographers seem to have overlooked Ben Joul, perhaps without due cause.

it. He thought no landscape half so fine as a long stretch of sand, flat and depressionless as is the sea, and with a stunted sandarac here and there, a few rare suddra bushes, and in the distance, an oasis, green as an emerald, with its wells, its melon fields, and clustering date palms, with their roots in water,* and feathery branches in the fire of the sun. No friend of French or English intervention was the Sherif Mulai Othmar; but a believer in the regeneration of the Moors, by a new intermixture of the desert blood which in times past has often been the salvation of the Arab race. From what he said the ancient Arab manners must have been preserved in all their purity in the Sahara, and, but for the introduction of villainous saltpetre, differ but little in essentials from a thousand years ago.

The low black tents of camels' hair, the wandering life, the little Arab saddles, used today in Syria and in Arabia, and differing widely from the high-peaked saddle of Morocco, the finer breed of horses, and, above all, the pure speech of the Koreish not mixed with Spanish and with Berber words as is the Arabic of the Morocco Moors, all show the great tenacity with which the desert tribes have clung to usages and to traditions hallowed by custom and by time. Finding the Sherif spoke such good Arabic, Lutaif and I agreed to refer to his decision a question of literary interest which had engrossed us for the past few days. A

---

* A palm tree is said by the Arabs to grow with its roots in water and its head in fire.

controversy at the time was raging in Morocco amongst the men of letters, as to whether the word *Mektub* (" It is written ") might not be spelt " Mektab."

Much had been written, as is usual in such cases, on either side, old friends had quarrelled, and each side looked upon the other as people of no culture and outside the pale of decent men. My preference was for " Mektub," but Lutaif, who had the prejudice which knowledge of a language sometimes induces, was in favour of " Mektab." The Sherif considered carefully, and gave his dictum that either could be said, although he added, those who say " Mektab " show want of education. So we were left in the position of the grammarian, whose last words were, " l'un et l'autre se disent."

Thus, in a week, I had met three Arabs all representative of their several classes, and, as usual, liked the man best who had been least influenced by European ways. Mulai Othmar disdained the idea of European protection, saying it was fit for Jews and slaves, but not for men, and that, for his part, sooner than ask for protection from any European power, he would return into the desert and follow a nomadic life. I took my leave of him with a feeling of real regret, such as one feels occasionally for those one sees but for an instant, but whose features never leave one's mind. If there are, in the Sahara, many such as he, there is regeneration yet for the Arab race, so that they resolutely refuse all dealings with Europeans; reject our bibles, guns, powder, and shoddy cottons, our political intrigues, and strive to

live after the rules their Prophet left them in his holy
book. If they forsake them, as in some measure the inhabitants of Morocco certainly have done, slavery, sure and
certain, is their lot; and in the time to come our rule, or
that of the French or Germans, will transform them into
the semblance of the abject creatures who once were free
as swallows, and who today lounge round the frontier
towns in North America calling themselves Utes, Black-
feet, Apaches, or what not, ghosts of their former selves,
sodden with whisky, blotched with the filthy ailments we
have introduced, and living contradictions of the morality
and the religion under which we live.

By noon next day we had almost dropped the Atlas out
of sight; the enormous wall of rocks rising straight from
the plain had vanished; the tall snowpeaks above the chain
alone remained in sight, and they appeared to hang suspended in the air. The vegetation changed, and once again
the ground grew sandy. The white* broom bursting into
flower covered it here and there in patches, as with an air
of snow new fallen and congealed upon the branches of
the plants. Again we passed a range of foothills, rocky and
steep, from the top of which, like a blue vapoury haze, we
saw the sea; and as we led our jaded animals down the
abrupt descent, a Berber shepherd standing on a knoll was
playing on his pipe. He stopped occasionally and burst
into a strange, wild song, quavering and fitful, the rhythm
interrupted curiously, so as to be almost incomprehensible

* Called Retam by the Arabs, and Retama by the Spaniards.

to ears accustomed to street organs, pianos, bands, sackbut, harp, psaltery, and all kinds of music which we have fashioned and take delight in according to our kind; but which I take it would be as void of meaning to a Berber as is our way of life. I check my horse, and sitting sideways for an instant, tried to catch the rhythm; but failed, perhaps because my ears were dulled by all the noises of our world, and less attuned to nature than those of the brown figure standing on the rock. But though I could not catch that which I aimed at, I still had pleasure in his song, for, as he sang, the noise of trains and omnibuses faded away; the smoky towns grew fainter; the rush, the hurry, and the commonness of modern life sank out of sight; and in their place I saw again the valley of the N'fiss, the giant Kasbah with its four truncated towers, the Kaid, his wounded horse, the Persian, and the strange entrancing half-feudal, half-Arcadian life, which to have seen but for a fortnight consoled me for my failure, and will remain with me a constant vision (seen in the mind, of course, as ghosts are seen); but ever fresh and unforgetable.

Next day about eleven o'clock, driving our horses through the Argan scrub in front of us, tired, dusty, and on foot, we reached the Palm-Tree House.

APPENDIX

Appendix A

" Some Observations on the Shillah Language."

THE Shillah tongue, that is, the speech of the southern branch of
the great Berber family, since 1799, when Venture de Paradis
made his vocabulary, has greatly interested students of philology.
Its great antiquity is undoubted, the term Amazirgh by which the
Shillah designate themselves occurring in the pages of several clas-
sical writers under the form of Mazyes, Mazisci, and Mazyes.
Though akin to the dialect spoken in the Riff mountains, and that
of the Kabyles of Algeria, and possessing considerable affinity to
the Tamashek spoken by the Tuaregs, it yet has considerable
dialetic differences from all of them, though they have been usually
held to be of one stock. The chief peculiarity that strikes a stranger
is the formation of the feminine in nouns which is marked by
the addition of a T at the beginning and the end of the word, as
Amazirgh (noble), which becomes Tamazirght, when used for the
speech of the Shillah people. Tarudant and Tafilet are merely
Arabic words turned into Shillah by the addition of the two T's,
and rendered feminine to agree with the Arabic word Medina, a
city.

The Shillah proper inhabit the whole range of the southern
Atlas, the province of Sus, with that of the Ha-Ha, and extend
southward to the oasis of Tafilet, though there Arab and Shillah
(Berber) tribes live close beside each other.

Place-names to which the word Ait (*i.e.,* Ben or Mac) is prefixed,
as Ait-Usi, Ait-Atta, Ait-M'tuga, etc., indicate that the district is
inhabited by a Berber tribe.

I subjoin a short list of words, written down whilst detained in Kintafi, but without pretending to their absolute correctness, as my knowledge of Arabic (our means of communication) is very slight, and nothing is more difficult than to get uneducated men to repeat words, familiar, and therefore easy to themselves, several times over, so that a stranger may catch them. In order to show the entire dissimilarity of the two languages, I give the Arabic as well as the English equivalents.

Tamazirght.	Arabic.	English.
Athghroum	Hobz	Bread
Aman	El ma	Water
Araras	T'rek	Road
Agmar	Aoud	Horse
Asardoun	Baralla	Mule
Ariyal	Hamar	Ass
Adrar	Gibel	Mountain
Anri	Bir	Well
Asif	Wad	River
Ergaz	Rajel	Man
Tamaghrat	M'raa	Woman
Afruch	Oueld	Son
Tafrucht	Bind	Daughter
Arroumi	Nazrani	Christian
Tagartilt	Hazira	Mat
Tifluth	Bab	Door
Imaguru	Sareuk	Robber
Whyh	Eiwah	Yes
Oho	Lawah	No

The language is, to my ear, not so guttural, but more nasal than the Arabic of Morocco. I think that a stranger, ignorant of both languages, would acquire Shillah more easily than Arabic. Like the Romany, or Calo, spoken in Spain, Shillah has been greatly corrupted by an admixture of the dominant tongue, so much so that the native verbs are largely lost, and their use supplemented by verbs taken from Arabic.

In writing, the Shillah use the Arab* character, and as far as is known, only one book, known as *El Maziri,* a compendium of the observances and ceremonies of the Mohammedan religion, has been written in the Shillah tongue.

The language abounds in the nasal gh, which may be rendered by an extremely nasal r, but which, like the Arabic h (Ha), is almost impossible to be learnt, except from someone acquainted with the proper pronunciation, as no written directions explain the peculiar sound.

Many treatises on Shillah, Kabyle, and Tamashek, and their differences and affinity, have appeared at different times, chiefly in French. Perhaps the best is that of the Marquis de Rochemontein, " Essai sur les rapports grammaticaux entre l'Egyptien et le Berbère," Paris, 1876. Leo Africanus also asserts the affinity of the two tongues.

Appendix B

THE author addressed the two following letters to the *Daily Chronicle* and the *Saturday Review* when detained at Kintafi. The freedom which he now enjoys having brought with it a form of mind more fitting to an ideal captive, he now doubts whether he would not have done better to have addressed his two letters to the *Record* and the *Rock.* Nevertheless he publishes his letters, hoping that an intelligent, and no doubt idealistic public will discern in them that resignation, trust in a higher power (as Turkey), hope, charity, or whatever is proper, that in such circumstances ought to have been found in such letters.

---

* The Tuaregs in writing Tamashek use a character of their own, but this is unknown to the Shillah.

"Thelata el Jacoub, Kintafi, Atlas Mountains,
"22nd October, 1897.

"The Editor, *Daily Chronicle*, London.

"Sir—It appears that like St. Paul I am destined to be in prisons oft. Whilst endeavouring to cross the Atlas into the almost unknown province of Sus, I was arrested by the Governor of this province on October 19th, and have been detained here on various pretexts ever since. Tonight one of our followers is to gird up his loins, tighten his turban, take his staff in his hand, pull up the heels of his shoes, testify to the existence of the one God, and strike across the hills to place this 'copy' in the hands of the British Vice-Consul at Mogador, about two hundred miles away.

"Though we are civilly treated, our position is the reverse of pleasant. We are allowed to walk about, but we cannot go far from our tent, and we have no idea why we are detained. I spare you any remarks on the flora and fauna of the district, for Inshallah, I propose to inflict them on a harmless and much book-ridden public. I merely state briefly that this house, an immense castle built of mud, is situated in an amphitheatre of hills, all capped with snow, and that a brawling river, the Wad N'fiss, runs past our tent; goats wander in the hills, tended by boys wild as their ancestors, whom Jugurtha led against the Romans. Horses and mules are driven down to drink by negro slaves, prisoners clank past in chains, knots of retainers armed with six-foot guns stroll about carelessly, pretending to guard the place; it is, in fact, Arcadia grafted on feudalism, or feudalism steeped in Arcadia. The call to prayers rises five times a day; Allah looks down, and we sit smoking cigarettes, waiting for you to turn your mighty lever on our behalf.

"For my companion in adversity I have a Syrian Christian, who acts as my interpreter, and who writes this for reasons known to you. Should Britain fail us, we hope that that great prince, the Sultan Abdul Hamid (God hath given him the victory), will send his fleet to our assistance, for, as we know, each of his Christian subjects is as a portion of his heart.

"Things look a little serious, as we are quite uncertain how long the Governor may keep us here. Therefore, I hope that this may go into your best edition, and be the means of making the Foreign Office act at once on our behalf, if we are not released.

"Yours faithfully,

"R. B. CUNNINGHAME GRAHAM.

"P. S.—Pray assure the public that we shall steadfastly refuse to abjure our faith."

Having thus done all in my power to invoke the protection of the Nonconformist Conscience (powerful amongst the noble Shillah race) for myself, and that of his Sultan for Lutaif, I recollected a business engagement, and wrote the following letter to excuse myself for non-completion of contract. I have ever held contracts as the most sacred of all the affairs of life.

"THELATA EL JACOUB, KINTAFI, ATLAS MOUNTAINS,

"22nd October, 1897.

"The Editor, *Saturday Review*.

"SIR—It will, I fear, be impossible for me to review the work called the *Canon*, about which I spoke to you. I hope, therefore, that you will place it in competent hands, for it is a well-written and curious book. You know that, as a general rule, I am reluctant to undertake reviewing, but in this case I should have been glad to make an exception to my usual practice.

"Before reviewing a book, I like to place a copy of it upon my table, and, after looking carefully at the outside of it, peruse the preface, glance at the title-page, read the last paragraph, and then fall to work. On this occasion, title and last paragraph, even the preface (which I understand is worthy of consideration), are beyond my reach. Not to be prolix, I may explain that for the last four days I have been a prisoner in the Atlas Mountain, at the above address, and that there seems no speedy prospect of my release. For details see the *Daily Chronicle*, to which I have addressed a letter, with one to our Ambassador at Tangier, which will, I hope, arrive some day, for when night falls our messenger is to endeavour to cross the hills to Mogador, our nearest post-town, some two hundred miles away, and to inform the Consul of our case.

"I am, Sir, yours faithfully,

"R. B. CUNNINGHAME GRAHAM."

Appendix C

THE following article appeared in the *Saturday Review,* and may serve to show one of the elements of difficulty against which I had to contend. Quite naturally, the country people thought that I was a filibuster.

THE VOYAGE OF THE "TOURMALINE."

The southern province of Morocco—that which extends from Aga-dir-Ighir to the Wad Nun—is called the Sus. Hanno is said to mention it in his famed Periplus. The Romans knew it vaguely. Suetonius may or may not refer to it when he speaks of a rich province below the Atlas; but his work is lost, and what remains comes down to us through Pliny, who himself laments the Romans took so little trouble to explore the coast. Polybius wrote of it; but what Polybius knew about the Sus is left so vague, that renowned, grave, arm-chair geographers have almost come to blows about it, as men of literature have done as to the whereabouts of Popering-at-the-Place.

Pliny certainly saw the lost writings of King Juba, and in them he met the word Asana. This Asana is conjectured (again by wise and reverend men) to have been "perhaps" Akassa, the Berber name of the Wad Nun. So that the ancients do not help us much to any knowledge of the Sus. Marmol and Leo Africanus talk of the province; but neither of them saw it, though Leo penetrated to Tamaglast, a village near Marakesh, supposed by some to be the hamlet now called Fruga. Thus little was known about the province, although travellers from Europe, as Arab writers tell us.

visited the capital Tarudant, coming by Agadir, during the sixteenth and seventeenth centuries, for purposes of trade.

All sort of legends thus sprang up about the place: demons inhabited it; a mountain spoke; magicians not a few lived near the
Wad Nun; La Caba, the daughter of Count Julian, who brought
the infidel to Spain, was buried in Tarudant, as the legend says;
and everything throughout North Africa, strange and miraculous,
occurred in Sus. Rich mines were there—gold, silver, and " diamont,"
iron, tin, and antimony, with manganese and copper; the people
were the most honest, wildest, wisest, and most ferocious in the
world; great ruined castles, known to the natives as " Kasbah el
Rumi," were dotted here and there, though who the " Romans "
were no one could tell, but probably they filled the place of the
" Moros," who, as is well known, built all houses, towers, and
buildings, of whatever nature, which exceed a hundred years in
age throughout all Spain.

Coming to more modern times, in 1791, the Sultan sent to the
Governor of Gibraltar for a doctor to cure his son, at that time
governor in the province of the Sus. An Army surgeon called
Lemprière was chosen, and, disembarking at Agadir, journeyed
to Tarudant. As far as anything is known, he is the first European
who entered Tarudant since the sixteenth century, when it is
certain that merchants from Holland used to journey to the annual
fair. He crossed the Atlas from Bibuan to Imintanut (by the same
pass, in fact, which I attempted in last October), and arrived in
safety at Mogador. He gives us little or no information about the
Sus, but vaguely speaks of mines, says that the country about Tarudant was fertile and well cultivated, and describes the pass he
crossed as skirting along tremendous precipices, which, to my certain knowledge, is not the case.

After him comes Jackson, who published his account of the
Empire of Morocco in 1809. Ball, in his appendix to Hooker's
" Morocco and the Great Atlas," refers to Jackson's book as being
the most copious ever written abut the Sus. Certainly he had special
advantages, for he passed sixteen years in Mogador, and Aga-

dir (now closed to trade), spoke Arabic and Shillah, but all he says does not amount to much. The map he made Ball considers inferior to that of Chenier, published a hundred years before his time. And so of Admiral Washington, Gerhard Rohlfs, Gatell, and Oskar Lenz. They all say little, for the good reason little is known. Although the last three travellers passed through the land, they went disguised, in terror of their lives, and are believed to have known little or no Arabic. So that it comes to this: All that we know with certainty is that a province called the Sus exists, that it stretches from close to Agadir to the Wad Nun, a distance of some two hundred miles, with a varying breadth of about seventy at the north, where it is bounded by the Atlas mountains, the Wad Sus, and the province of the Ha-Ha to a hundred or more at the extreme south, as no one knows how far the boundaries of the province stretch up the waters of the river Nun.

Round about Agadir the country has been visited, and is reported to be very like the provinces of Shiadma and the Ha-Ha, which bound it to the north, that is, it is in general configuration flat and sandy, with stretches here and there of reddish argillaceous soil, but both soils greatly grown over with thorny bushes, and here and there well cultivated. Politically the province owns the Sultan of Morocco's sway, but his authority extended lately but to Tarudant, the district called Taseruelt, in which is situated the Zowia of Si Hamed O'Musa, now represented by Sidi Haschem, and to the great Arab tribe of the Howara who occupy the country between Fonti and Tarudant. Up to the banks of the Wad Nun, where there are Arab tribes again (but wild and independent of the Sultan), most of the inhabitants of the country are of the Berber race. This race, the original inhabitants of the country before the Arab conquest, has never been entirely conquered, and between them and the Arab conquerors a strong enmity exists.

The chief trade of the province has always been with Mogador since the port of Agadir was closed by the great-grandfather of the present Sultan. It consists of wool and camels' hair, goat-skins and hides, bees' wax, a little gold dust, ostrich feathers, gum-arabic, cattle and sheep, almonds, and all the products of the Sahara, for most of the trade from the western portion of that district comes to Mogador. In return, they take Manchester goods, powder, tea, sugar, cheap German cutlery, and all the wonders which human nature has to suffer to produce, and enrich the manufacturers of Leeds, Manchester, Birmingham, Liège, Roubaix, and the like in turning out. So thus the situation briefly stood.

A province, large and wealthy, the mouth of trade with the Sahara, supposed to contain rich mines, though on this head nothing is known with certainty, except that a little copper is worked near Tarudant, though the natives say gold, silver, iron, and magnetic ironstone exists; fertile in climate, thickly populated, and ill affected to its ruler; fanatical and largely swayed by a sort of general Witenagemot known as the "Council of the Forty," and yet the population bound to get all supplies of European goods through the one port of Mogador.

Many and various have been the attempts to open trade direct. Pirates and filibusters, and traders with a moral sense of what was due to civilization and to themselves, had all attempted many times to supply the poor heathen with their European trash, but never with success. Sometimes they landed, were taken prisoners, and a "diplomatic question" was superinduced until they were released. At other times they disappeared on landing and were never heard of, but still reports poured into Mogador of the great riches of the Sus. These riches to my mind are non-existent, for I have known hundreds of Susi traders, merchants, camel-drivers, tribesmen, "saints" and acrobats, from Taseruelt, but never saw a Susi who was rich.

In general, I found them tall, thin, dark-coloured men, very intelligent, fanatical, great travellers, petty traders; now and then ostrich hunters, and sometimes slave-dealers, but all were poor, although when asked they always talked about gold-mines and the riches of their land, and showed an evident desire that the various ports along the coast should be left open for European trade.

Then came the death, about four years ago, of the late Sultan

Mulai el Hassan (may God have pardoned him!), and the disturbances consequent on the accession of a minor to the throne. The Susis without doubt thought the time suitable for movement, and no doubt hoped to be independent, and to buy powder, tea, and sugar and cotton goods, without the trouble of coming up to Mogador. Rebellions more or less partial took place throughout the province, and were subdued.

About two years ago, in Mogador, appeared one Captain Geyling, a Jewish Austrian subject, who, by some means or another, got into communication with certain discontented chiefs, whom he induced to sign a treaty with him to open up a port, start trade with the interior, work the mines, and generally to allow the country to be brought under the humanizing influence of European trade. This done, he straight repaired to London, and tried to form a company, but found out, as Lydgate did before him; that " lacking money he mighte never speede."

Then came a hiatus, which perhaps some of the gentlemen who planked their money down may like to fill up for the benefit of those who take an interest in unofficial efforts to extend the shadow of our flag.

We next find Captain Geyling back in Morocco, dressed in a single-breasted black frock-coat and fez, and turned in the interval into a pseudo-Turk under the title of Abdul Kerim Bey. Here history says, as the advance agent of the Globe Venture Syndicate he travelled like a prince, taking as many tents as would befit a travelling menagerie, plate and more plate, servants and horses, mules, guns, presents for the Kaids, and impelled by a consuming thirst to get concessions for his paymasters. With him as military adviser, attaché, or what not, went Major Spilsbury, and why he let himself be towed about the place by Geyling only he can tell.

Quiet but determined, a linguist, leader of men, and one of those willing to risk his life ten times a day for any syndicate, upon most reasonable terms, Spilsbury was a born filibuster. Always about to make a fortune with schemes innumerable, in which if you embarked you still stayed poor, or became poorer; but with this difference from the schemes of most men of his class, that he himself was never richer by a penny by any one of them. Geyling had said he knew the Sultan well, and as that potentate was somewhere in the south, Geyling proceeded north and waited upon the Sherif of Wazan, the spiritual head of all things in Morocco. There he seems not to have had much luck, and then went east to Fez, back to the coast, and after two or three months' perambulation up and down the land, went to Morocco city, where he ought to have gone first. There neither Sultan nor Vizir would see him, and with his tail between his legs, he returned to Mogador, and in a little inn kept by a Jew quarrelled with Spilsbury, who, if reports be true, threatened to beat him with a stirrup leather, and the companionship broke up. Geyling Kerim went homewards to Vienna, Novi Bazaar, or for all I know joined his repatriated co-religionists in their new colony in Palestine. But Spilsbury, being apparently determined to play things out "on a lone hand," remained in Mogador, and then embarked upon a series of adventures, the more extraordinary when one remembers that he speaks no Arabic.

How, wherefore, in what manner, or by what means, he came across him I do not know, but he fell in with an acquaintance of my own, one Mr. Ratto, born in Mogador, and speaking Arabic, and Shillah, French, English, Spanish and apparently all other tongues with equal ease. What actually they did, only themselves are in a position to record, but I suppose that, taking advantage of the unsettled state of things, the Sultan's absence punishing refractory tribes, and the desire which every Arab chief has of getting arms to make himself quite independent of all mankind, they must have entered into negotiations with some of the chiefs of the wild tribes in Sus. Spilsbury seems to have satisfied the Syndicate in London that they could trade direct with Sus, receive concessions from the chiefs, land and construct a factory, and in time make themselves sole masters of the place. No doubt they reasoned: If we are once established, when troubles come, England must for her honour protect her subjects, and in protecting them, protect their interests, and they knew that England once committed to interference in any country (said to be rich), must of necessity remain to restore order, introduce good government, and generally to further the cause of progress and morality, which is specially her aim in every country peopled by an inferior race. What treaty Spilsbury took home is matter of conjecture, but not unlikely he got signatures from chiefs, who signed, thinking if all went well they would gain something, and if things turned out badly they could say they had been deceived, and signed a document that they had not understood. One name is certain was appended to the deed, that of M'barek-ou-Ahmed, who is now securely chained in some pestilential prison in Fez or Mequinez. Be all that as it may, Spilsbury was shortly back again in Mogador, trying to hire a vessel to convey himself, a Jew interpreter, and several samples of. his goods down to Akssis, a port between Wad Nun and Agadir. But by this time the Sultan had got wind of the affair, and sent his emissaries into the Sus to bribe the chiefs into allegiance, and what is more he had communicated with the English Ambassador in Tangier, who having sent the news to London, the expedition and its aim was laid before the Foreign Office. Presently an official notice appeared declaring that the British Government viewed with concern the meditated attempt to open trade with a part of the Emperor of Morocco's territory against his will, and if any person went for such a purpose, he must go at his own risk. Spilsbury probably cared nothing for the protection or the displeasure of either Government, so he pushed on his preparations just as if nothing had occurred worth mentioning.

The ukase of the British Government had made it difficult to operate from Mogador, but Spilsbury, nothing dismayed, engaged a Jew interpreter, and all alone, or at the most with two or three companions, sailed for the Canaries, hired or bought a schooner, and after a passage of an abnormal length, contending all the time with contrary winds, sailed to Akssis, landed, and started to palaver with the chiefs who were expecting him, with several thousand men encamped upon the shore, having been warned most probably by Mr. Ratto to hold themselves in readiness against his coming.

Nothing more different than the inception of the Jameson affair and that so boldly planned by Spilsbury. Both gentlemen adventurers, or if you like, both advance agents of the British Empire. One flag-wagging and backed up by all the " fruit secs " of the British army, champagne and sandwiches laid on at every twenty miles upon the road; the other almost alone upon a coast not visited twice in a century by Europeans, and in the hands of men who kill a man with as few compunctions as a settler up in North Queensland flogs a black to death. If he had goods to sell I know not, if he had samples of trade powder and trade guns, that is to me unknown, but anyhow, by the assistance of his interpreter, he entered into a council with certain of the chiefs, as the Sherif of Taseruelt, the aforenamed M'barek-ou-Ahmed, and others whom it is better not to name, and was about to sign a treaty with them, to open trade direct, put up a factory, work the mines, and generally prepare the way before the faces of the Globe Venture Syndicate.

But for an accident Spilsbury might have been Emperor of Agadir, the Lord Protector of the Sus, or Rajah of Tamagrut, but Fate or the Sultan of Morocco had otherwise disposed.

Most of the chiefs of Sus were at Akssis with many of their followers, but one Sheikh with about fifty horsemen had kept aloof during the progress of the negotiations, either because he had not been considered big enough to square, or as some think because he was secretly acting under orders from the Moorish Court.

Just as the chiefs were about to sign, and each one had agreed how many rifles he was to receive on Spilsbury's return with a well-laden ship, the Sheikh mounted his horse, marshalled his followers and plunged into the middle of the crowd, yelling and firing several shots, exclaiming: " Out with the Christians ! I will not be a party to any dealings with our hereditary foes." Thinking they were attacked in force, the followers of the other chiefs returned the fire, all was confusion, and Spilsbury, to save his life, had to retreat precipitately on board his ship, and to complete the scene, the smoke of a steamboat was seen coming down the coast. Now, as no vessels between Agadir and the Wad Nun come near the coast, which is one of the most deserted in the world, Spilsbury knew at once it must be a vessel of the Moorish Government upon the search for him. Luckily night was near, and a fair wind sprang up which took him to the Canaries, from whence he shipped aboard a steamer and returned to England to plan another trip.

What actually he did in England during the next six months I do not know, but in November I met him at a London club, the proud possessor of the steam yacht " Tourmaline," carrying a quick-firing gun, an assorted cargo of goods fit for the Morocco trade, and some nine thousand rifles with which he intended to arm his friends, the followers of Sidi Haschem, and the other Shiekhs of Sus. The vessel lay at Greenhithe, and was to sail next morning for Antwerp to take the rifles in; yet Spilsbury sat smoking quietly without a trace of " Union Jackism," no word of " moral purpose," not a suggestion of being, as Dr. Jameson seemed to think he was, a sort of John of Leyden going to set a people free. Simply an ordinary club man, talking of what he was about to do, as he had talked of fishing in Loch Tay. A well-dressed, quiet-mannered filibuster, not bellowing that he would make the Arabic language popular in Hell, after the " fighting Bob Tammany " style, but quite aware that he was venturing his life, and perilling for ever such reputation as he had. As a law-abiding citizen, I tried to show him all the error of his ways, spoke of the wickedness of all he was about to do, and watched him get into a cab with mingled feelings of disgust at the peddling syndicate which, for its wretched five per cent, was about to bring the name of England into contempt, and admiration for the man who was going quietly to risk his life in such a miserable cause.

How he sailed, reached Akssis, landed some rifles, was interrupted in his dealings by the arrival on the one hand of the Sultan's troops under Kaid el Giluli, and on the other by the advent of the Moorish Government's armed transport, " El Hassani "; how he exchanged shots with her, rescued his boat, but failed to save his four companions who remained captives; and how he with the yacht " Tourmaline " is still detained, or was up to the other day, under surveillance in Gibraltar, is well known to all. But what befel his four companions and the unfortunate M'barek-ou-Ahmed, his intermediary, has never been made public in this country yet. So, to make matters plain, I quote the letter of a French Algerian gentleman settled in Mogador:—

" Vous me demandez des nouvelles de The Globe Venture Cie., c'est bien une aventure que les bondholders Anglais ne goberont pas facilement. Vous savez que le commandant en chef, Major Spilsbury, est arrivé à Akssis entre Agadir et Wad Nun, et là il a débarqué 500 fusils, 100 caisses cartouches, 4 balles cotonnades, 25 caisses thé, etc.

" Le Major a eu le bon esprit de rester à bord, mais a fait débarquer un jeune Anglais, le second du bord, un allemand comptable, un Juif interprète, et un marin portugais. Voilà un bouillabaisse! Enfin ils se sont établi sous trois tentes sur la plage, avec le Sheikh avec qui le Major avait fait connaissance chez M. Pepe Ratto, et huit ou dix Arabes de l'endroit. Après deux jours la frégate ' El Hassani ' de sa Majesté Chérifienne est arrivée, en même temps le Kaid El Giluli arrive par terre avec 500 cavaliers, entoure les tentes et toute la boutique est prise. Il y a eu quelque coups de fusils, deux hommes du Major (Arabes) sont blessés, la Tourmaline s'éloigna vers Lanzarote, les quatre Européens et le Juif sont pris, aussi que le chef Arabe (M'barek-ou-Ahmed) avec vingt de ses amis, et l'aventure est fini.

" Les Arabes ont été conduit au Sultan qui les a envoyé avec chaines se pourrir dans les prisons de Fez, et les quatre Européens sont depuis près de trois mois dans la maison du Kaid el Giluli à Ha-Ha, a savourer la Shisha (vous en connaissez le gout oh Sheikh Mohammed el Fasi), en attendant les ordres de Sidna. . . . Le pauvre Sheikh, M'barek-ou-Ahmed, n'avait aucune influence sur les tribus, et il ne marcha que sur la promesse que les ' fregatas Inglise ' viendront débarquer des soldats anglais, le pays sera pris, on fera du Sheikh un Kaid, et cela accompagné de cent dollars, 20 livres de thé, et un sac de sucre, et le pauvre Sheikh a eu l'eau (et le thé) à la bouche.

" Un soldat qui avait conduit les prisonniers au Sultan, a dit, que sur la route, le pauvre Sheikh disait tout le temps, ' Oh, le Nazrani

m'a trompé, ce sont des trompeurs les chrétiens, il m'avait promis que
des frégates et des soldats anglais avec des canons débarqueront, et
aux premier coup de fusil, son bateau s'est sauvé; par Dieu, ce
chrétien doit être un Juif! mais c'était écrit, Allah Ackbar.' "

"R. B. Cunninghame Graham.